Biophysical Osteoblast Stimulation for Bone Grafting and Regeneration

Nahum Rosenberg

Biophysical Osteoblast Stimulation for Bone Grafting and Regeneration

From Basic Science to Clinical Applications

Springer

Nahum Rosenberg
Department of Research and Development
Sheltagen Medical Ltd
Atlit, Israel

ISBN 978-3-031-06922-2 ISBN 978-3-031-06920-8 (eBook)
https://doi.org/10.1007/978-3-031-06920-8

This Springer imprint is published by the registered company Springer Nature Switzerland AG
The registered company address is: Gewerbestrasse 11, 6330 Cham, Switzerland

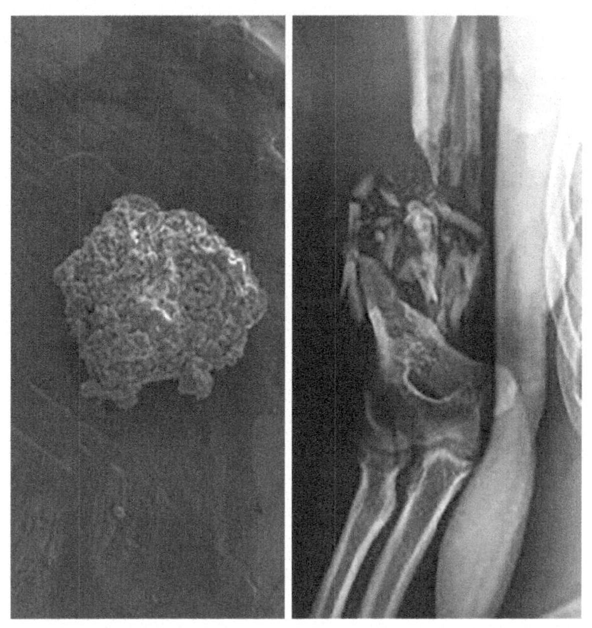

Preface

Facts relating to attempted osseous transplantations are few, the details concerning them are meagre, and the deductions drawn from them have been received with dubiety.

Sir William Macewen. Observations concerning transplantation of bone. Illustrated by a case of inter-human osseous transplantation, whereby over two-thirds of the shaft of a humerus was restored. Proceedings of the Royal Society of London 1981; 32:212–215.

This book, partially based on my Ph.D. thesis,[1] provides an overview of the fascinating ability of bone to re-generate under external biophysical control. The clinical importance of this phenomenon is apparent because human skeletal support can recover after local or systemic damage. The main cellular component that governs this response is the osteoblast, which is highly susceptible to external biophysical stimulation.

This is not an encyclopedic text but rather an attempt to present the interrelation of cellular mechanisms with the potential to be tuned for the clinical application of different biophysical energies for bone regeneration by stimulating osteoblast metabolism.

Therefore, this text should help surgeons to understand the cellular mechanisms that underline the various currently used or predicted clinical methods for enhanced bone regeneration by biophysical methods.

Scientists may benefit from recognizing the clinical goals of the basic research into the biophysical stimulation of osteoblasts.

These mutual benefits are the primary goals of this book.

[1] Nahum Rosenberg, "High-frequency alternating biophysical stimulation of human osteoblast" (School of Pharmacy and Biomedical Sciences, Faculty of Science, University of Portsmouth, October 2021).

Historic Insights

The ability of bone to regenerate has been recognized by man since ancient times. Additionally, it has always been well known that the prerequisite for successful bone healing following a fracture is the provision of adequate external physical effects, such as mechanical splinting of the fractured site.

There is evidence from archaeological excavations that ancient Egyptians, even before 1075 BCE, used wooden splints for limbs with fractured bones with some success for fracture healing, especially in the upper limbs [1]. This relatively nonsophisticated method for healing of a bone gap (fracture) by solid external splinting was practiced by different cultures on different continents. It was practiced by Greeks and Romans (evidence from around 400 BCE), seventh century CE Byzantines [2], and fifteenth century CE Aztecs, who skillfully splinted the fracture by external "plaster-like" splints and even by intramedullary wooden nails [3].

Furthermore, the apparent ability of bone to regenerate and close the gaps was utilized even in the Stone Age. There is evidence from 2000 BCE, in the excavation of the ancient Khurits community, of successful implantation of xenograft into calvarial traumatic 6-mm gap in 35-year-old women. This xenograft incorporated well and bridged the gap [4]. Because the concept of the possibility of using xenograft as a bone graft has a clear, intuitive basis, it has not been abandoned. An example is a successful cranial gap bridging following a bone xenograft implantation originating from a dog into a cranial defect of an injured soldier. This procedure was performed by Job van Meekeren (Dutch surgeon) in 1668 [5].

Eventually, on the stepway of modern surgery, three main essential components for a successful bone regeneration process were recognized, i.e., the optimal mechanical environment of the bone gap site and adequate local vascularization and blood supply. According to these basic concepts, additional attempts at bone grafting have evolved in those three main directions by using xenografts, allografts, and autografts.

Before the immune rejection trait was recognized, xenografts were apparently the most appealing source of bone graft material. In 1891, an ambitious procedure of implantation of vascularized bone xenograft into non-union of the fractured tibia in two patients was performed by A.M. Phelps at the Charity hospital in New York. In this procedure the graft remained connected to the blood supply of the donor dog, aiming to provide an adequate blood supply to the graft. In both cases the procedure failed because of insufficient mechanical stabilization and probably owing to immune rejection [6].

The failure to use xenografts evolved into bone grafting by allografts. There is a documented report of allograft implantation, originating from three donors, into a large bone gap in the humerus (two thirds of the humeral length) of a 3-year-old child following resection owing to osteomyelitis. The procedure was performed at the Royal Infirmary, Glasgow, by Dr. William Macewen in three stages during 1879–1880. The allografts were obtained following osteotomies in three children who were treated for tibial bowing due to rickets. By this procedure, almost the entire length of a viable humerus with good vascularization was restored [7]. This phenomenal surgical success was not widely supported by other reports on implantations of bone allografts; therefore, the autograft was the most conceptually appealing method for bone grafting. The initial documented surgical attempt at bone autografting was made in 1820 by Philips von Walter [8] to close a calvarial bone defect.

In addition to the development of more efficient methods of bone grafting, the well-known concept of the provision of an optimal physical environment for bone regeneration contributed to the understanding that any bone regeneration technique should take into consideration these two main aspects—tissue viability and mechanical environment. This concept was consolidated in 1892 by Julius Wolff from the University of Berlin, who, following extensive research of bone specimens with the support of mathematical modeling, concluded that bone regenerates according to the vectors of external mechanical forces (Wolff's Law) [9]. This understanding is the basis of the future evolution of the concept of mechanotransduction in bone-generating cells (osteoblasts).

With the further development of surgical techniques at the start of the twentieth century the consolidation of the evolved basic concepts of bone grafting was presented by the fundamental work of Fred H. Albee from the University of Vermont, who described bone-grafting surgical methods in the limbs and the spine [10].

The culminating method of bone regeneration induction techniques, following a strain effect generated by the externally applied force, is the "distraction osteogenesis" effect, which was popularized by the pioneering work of Professor Gavriil Abramovich Ilizarov in the mid-1950s [11]. This clinical method supported the theoretical concept of Wolff's law by showing that viable bone is generated owing to the locally applied optimal mechanical strain on bone following cortectomy.

About 10 years later additional mechanism that can enhance bone regeneration was gradually discovered and popularized by Professor Marshal R. Urist from the University of California. The bone morphogenic protein (BMP) group of cytokines showed a pronounced osteogenic effect in vivo with the ability to bridge bone defects [12]. Later, it was discovered that the cellular pathways related to BMPs are part of the mechanotransduction process in osteoblasts [13], indicating the mechanical and humoral interconnections in the bone regeneration process.

The possibility of the induction of in vivo bone regeneration for the closure of critical gaps by using mesenchymal stem cells (MSCs) seeded on osteoconductive support began to gain interest following the development of effective techniques of MSC expansion in vitro [14]. The source of autologous MSCs can be bone marrow, adipose tissue, peripheral blood, and other mesenchymal sources. There is clinical evidence for the successful closure of the calvarial bone defect in a 7-year-old girl by in situ application of adipose-derived MSCs with the support of autologous fibrin glue [15]. This approach can be defined as "local cell therapy" for bone regeneration. The method has not gained widespread clinical application because the fate of the implanted progenitor cells cannot be predicted, as their osteogenic ability, following maturation to osteoblasts, which is governed by the local biomechanical vascularization factors, cannot be sufficiently controlled externally.

Table 1 General historical milestones of methods of bone regeneration

Epoch/ Date	Community/Surgeon/Scientist	Method
2000 BCE	Stone age Khurits community	Xenograft
1075 BCE	Ancient Egyptians	Splinting
400 BCE	Greeks	Splinting
700 CE	Byzantines	Splinting
1500	Aztecs	Splinting
1668	Job van Meekeren	Xenograft
1820	Philips von Walter	Autograft
1880	William Macewen	Allograft
1892	Julius Wolff	Wolff's law
1915	Fred H. Albee	Fundamentals of bone grafting
1954	Gavriil Abramovich Ilizarov	Distraction osteogenesis
1960	Marshal R. Urist	Bone morphogenic protein
2004	Justus-Liebig-University Medical School, Giessen, Germany	Cell therapy
Future	In vitro viable bone generation by tissue engineering	

The naturally evolving didactic question should be how the ability of bone to regenerate in vivo, which is orchestrated by the osteoblasts under biophysical control, can be utilized to generate live bone tissue in vitro for subsequential clinical use for bone grafting. There is initial evidence that viable bone tissue can be generated in vitro and act as an autologous bone graft in vivo following the exposure of MSCs to the optimal biophysical milieu. This manuscript deals with this intriguing subject (Table 1).

References

1. Smith GE. The most ancient splints. Br Med J. 1908;1(2465):732–6.2. https://doi.org/10.1136/bmj.1.2465.732.
2. Brorson S. Management of fractures of the humerus in Ancient Egypt, Greece, and Rome: an historical review. Clin Orthop Relat Res. 2009;467(7):1907–14. https://doi.org/10.1007/s11999-008-0612-x.
3. de Montellano BO. Aztec medicine and health, and nutrition. New Bruswick: Rutgers University Press; 1990. p. 149, 173, 174.
4. Noorbergen R. Secrets of the lost races: new discoveries of advanced technology in ancient civilizations. Teach Services; 2014. p. 170–1.
5. de Boer HH. The history of bone grafts. Clin Orthop Relat Res. 1988;(226):292–8.
6. Phelps AM. Transplantation of tissue from lower animals to man. New York: Trow's Printing and Bookbinding; 1891. p. 2–22.
7. Macewen W. Observations concerning transplantation of bone. Illustrated by a case of inter-human osseous transplantation, whereby over two-thirds of the shaft of a humerus was restored. Proc R Soc London Ser. 1881; 32:232–46.

8. Walter PH. Wiedereinheiling der bei der Transpanation ausgebohrton Knochenscheibe. J Chir Augen Heilkund. 1821;2:571.
9. Wolff J. The law of bone remodeling. Berlin: Springer; 1986 (translation of the German 1892 edition).
10. Albee FH. Bone-graft surgery. Philadelphia: W.B. Saunders Company; 1915.
11. Spiegelberg B, Parratt T, Dheerendra SK, Khan WS, Jennings R, Marsh DR. Ilizarov principles of deformity correction. Ann R Coll Surg Engl. 2010; 92(2):101–5.
12. Urist MR. The search for and the discovery of bone morphogenic protein (BMP). In: Urist MR, O'Connor BT, Burwell RG, editors. Bone grafts, derivatives & substitutes. Butterworth-Heinemann; 1994. p. 315–62.
13. Kopf J, Petersen A, Duda GN, Knaus P. BMP2 and mechanical loading cooperatively regulate immediate early signalling events in the BMP pathway. BMC Biol. 2012;10:37. https://doi.org/10.1186/1741-7007-10-37.
14. Gamie Z, Tran GT, Vyzas G, Korres N, Heliotis M, Mantalaris A, Tsiridis E. Stem cells combined with bone graft substitutes in skeletal tissue engineering. Expert Opin Biol Ther. 2012;12(6):713–29.
15. Lendeckel S, Jödicke A, Christophis P, Heidinger K, Wolff J, Fraser JK, Hedrick MH, Berthold L, Howaldt HP. Autologous stem cells (adipose) and fibrin glue used to treat widespread traumatic calvarial defects: case report. J Craniomaxillofac Surg. 2004;32(6):370–3.

Atlit, Israel Nahum Rosenberg

Contents

About the Author

Nahum Rosenberg MD, Ph.D., MOrthop (Hons.), MBA, FRCS (England), Consultant Orthopedic surgeon with more than 30 years of professional experience.

Dr. Rosenberg earned his MD degree from the Faculty of Medicine at Technion, IIT, in **1990**. He accomplished a clinical residency in Orthopedic Surgery at Rambam Medical Center, Haifa, in **1997**. He was extensively involved in basic research, which led to the MOrthop (Hons) degree from Tel Aviv University in **1996**. In **1998–1999** he was appointed as a Nuffield Fellow in Orthopedic Surgery at Oxford University UK, and in **1999** as a clinical fellow in orthopedics at Mercy Private Hospital, Melbourne, Australia. Then, in **2002**, he was appointed as a Clinical Fellow in Orthopedic Surgery (upper limb surgery) at the University of Nottingham, UK. In **2018**, he achieved an MBA degree with a specialization in Biomedicine at the College of Management, Academic Studies, **Israel.**

Since **2003** he has served as a senior orthopedic surgeon at Rambam Health Care Campus, Haifa. He was appointed Assistant Clinical Professor at the Faculty of Medicine, Technion, IIT, in **2007**. His scientific activity concentrated on basic and clinical research into bone tissue regeneration, which led to his Ph.**D.** degree at the University of Portsmo**uth, UK.**

Dr. Rosenberg is an active member of the editorial boards of **12** scientific journals. He has served as a member of the Shoulder Committee of ISAKOS (International Society of Arthroscopy, Knee Surgery and Orthopedic Sports Medicine), authored **80** peer-reviewed scientific publications, edited four books on orthopedic surgery and related basic science, and has authored **150** presentations at international scientific m**eetings.**

His main research interests are osteoblast physiology, bone biology, bone regeneration, shoulder surgery, the outcome of orthopedic procedures, and human biomechanics. His primary clinical expertise is in shoulder **surgery.**

Part I

In Vitro Bone Generation by Biophysical Stimulation. Prospective for Autologous Bone Grafting

The Theoretical Context of Biophysical Stimulation of Osteoblasts

1

Electromagnetic energy affects mammalian cells' phenotypic function and proliferative activity by altering transmembrane ion flux [1–3]. This response can be induced by mechanotransduction or another external energy effect (for example, by a direct electromagnetic field application) by changing the cell membrane potential. Electrical currents via cell membranes cause the activation of intracellular synthetic pathways. When stimulated by a flux of calcium and phosphate ions, the transmembrane currents generate electromagnetic fields that produce a phenotypic cell function even at extremely low intensities of 0.02–0.03 mT [1–3]. This intensity range of local electromagnetic fields (EMF) is three levels of magnitude higher than the range of the background magnetic field on Earth, i.e., 0.0003–0.0007 mT according to geographical location [4].

There are seven known subgroups of the Transient Receptor Potential (TRP) membrane channels, which are permeable for calcium ions. Member number 4 of the TRP subfamily V (TRPV4) is involved in mechanotransduction and osteoblastic commitment in the mesenchymal stem cells (MSCs) following external oscillatory mechanical stimulation. This channel is predominantly situated in the high-strain loci on the osteoblast cell membrane, i.e., adjacent to the focal adhesions and the primary cilium [5].

An additional membrane channel, permeable to the calcium ions and involved in mechanotransduction in osteoblasts, is PIEZO1. This channel is part of the mechanotransduction pathway induced by static pressure and shear stress that causes membrane stretch. PIEZO1 regulates the expression of collagens II and IX, which are part of osteoclast differentiation machinery [6]. Cellular mechanotransduction involves numerous cellular pathways propagating mechanical force from the extracellular milieu and translating it to the synthetic cellular response. The exact interrelation of these pathways is not sufficiently clear and is under extensive research efforts. Overall, this process is initiated by the external mechanical force causing shear stress on the cellular membrane and dictated by the mechanical stiffness properties of the supporting extracellular matrix. Cell biomechanics represent the

N. Rosenberg, *Biophysical Osteoblast Stimulation for Bone Grafting and Regeneration*, https://doi.org/10.1007/978-3-031-06920-8_1

interactions of different cell components with different mechanical characteristics, e.g., relatively stiff cytoskeleton vs. more elastic cell membrane. In general, cytoskeletal components provide a degree of rigidity that reinforces the highly elastic membrane, and this interaction determines the cellular shape and structural integrity in its mechanical environment. Thus, cell deformation is caused by two main mechanisms: uneven bending of the lipid components of the bilayer cell membrane and nonuniform contraction of the cytoskeletal components.

The deformation of the cellular membrane affects the cell membrane potential by inducing electrical currents [1–3] due to a piezoelectric effect when mechanical stress on the cell accumulates as an electric charge [7]. Additionally, the external force is transferred via transmembrane molecular extensions (Integrins, Cadherins) and cytoskeletal components to the nuclear lamins (predominantly lamins A and C) with a subsequential induction of transcription activity (Fig. 1.1) [10].

The external mechanical force is transferred to the nuclear lamins via perinuclear actin fibrils adjacent to the outer nuclear membrane. The lamins, which have elastic characteristics and are concentrated in the outer and inner nuclear membrane boundaries, propagate the mechanical force to the intranuclear chromatin through the proteins of the linker cytoskeleton complexes (LINC). Eventually, the force reaches the chromatin, which responds with a "spring-like" folding-unfolding activity. The unfolded chromatin reacts with a transcription of mechanosensitive transcription factors, e.g., MAL-SRF, and YAP/TAZ (Fig. 1.1b) [8].

Furthermore, mechanotransduction involves a transmembrane activation of the canonical WNT pathway, the receptors of Insulin-Like Growth Factor (IGF) and TGF-β, that eventually activate Osterix (Fig. 1.2). This transcription factor is

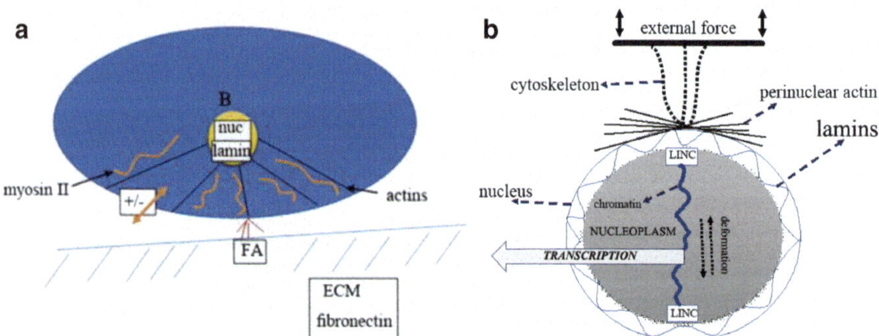

Fig. 1.1 Simplified schematic representation of the general mechanotransduction pathway. (**a**) The mechanical force propagates from the extracellular matrix (ECM) via focal adhesion (FA) sites consisting of transmembrane molecules, e.g., integrins, through cytoskeletal components to the cell nucleus (nuc) and eventually causes transcriptional activity (B). (**b**) External mechanical force propagates through the cytoskeleton via perinuclear actin to lamins (predominantly lamins A and C), and via the linker cytoskeleton complex (LINC) proteins, causing reversible deformation in chromatin with its subsequential transcriptional response. In parallel, mechanically induced cell deformation affects transmembrane ion flux (±), which is also responsible for the induction of cellular biochemical pathways [8, 9] (figure source—PhD Thesis authored by Rosenberg Nahum, "High-frequency alternating biophysical stimulation of human osteoblast," School of Pharmacy and Biomedical Sciences, Faculty of Science, University of Portsmouth, October 2021)

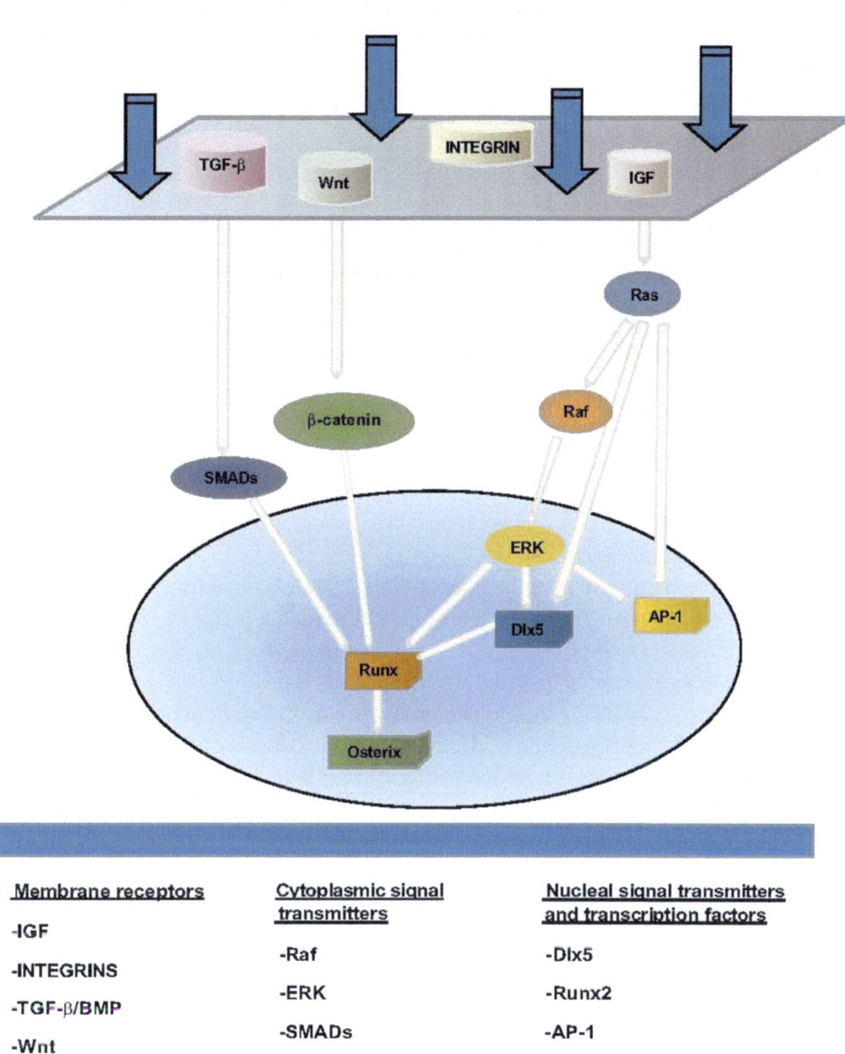

Fig. 1.2 Schematic representation of several biochemical pathways induced by mechanotransduction for activation of osteoblast maturation [9]. From "Berkovich Y, Shapira J, Rosenberg N. Chapter 5. Mechanotransduction in osteoblasts. In: Experimental methods for mesenchymal cell activation by biomechanical stimulation. Rosenberg N Ed. Bentham Science Publishers, 2016, pp. 49–57" with permission from Bentham

involved in osteoblast differentiation [11]. This process involves the RUNX 1/2 transcription factors that bind directly to DNA and, by interconnections with other osteogenic pathways (WNT/β-catetin, BMP), induces osteogenesis [12].

Therefore, there is a potential for the ability to manipulate mammalian cells' metabolism by external biophysical stimulation that alters the milieu EMF and subsequentially translates into specific intracellular biochemical pathways. The biophysical means for this purpose can be mechanical or electromagnetic, which, when applied in a specific energetic range and mode, causes cellular activation. This phenomenon is especially prominent in mesenchymal origin cells responsible for the mechanical support of the whole body, which readily responds to an external biophysical environment (Fig. 1.2) [9, 13].

Because the bioenergetics of osteoblasts has not been clarified sufficiently yet, even less known on the interrelations between the biophysical stimulation of osteoblasts and their cellular energy supply and demand.

The combination of metabolic and signaling pathways is necessary for the exploitation of intracellular metabolic resources and cell survival, proliferation, and differentiation. Depending on their role and environment, several energy metabolism phenotypes are displayed in different cells. Therefore, the stage and direction of differentiation have a significant impact on the catabolic pathway of glucose after it enters the osteoblasts. Aerobic glycolysis was found to an important and unique for these normal cell pathways in ATP generation (Warburg effect). Aerobic glycolysis can produce ATP more quickly, intermediate metabolites that are required to make matrix proteins, and secretes citrate, which is crucial for the formation of the inorganic structure in bone [14].

According to the in vitro studies, osteoblasts' glycolytic activity does not differ from that of undifferentiated cells. Still, there is evidence from in vivo studies that more mature osteoblasts exhibit higher oxidative phosphorylation activity [15].

Since the energy supply for mature osteoblasts depends on glycolysis, glycogenic amino acids constitute the raw material for the differentiated cells. The tricarboxylic acid cycle intermediates or pyruvic acid is produced during the metabolism of glycogenic amino acids, translocated to the mitochondria, and cause boosted expression of mRNA for the enzymes alkaline phosphatase, runt-related transcription factor 2 (Runx2), and osteocalcin, which is involved in osteogenesis. Naturally, fatty acids and glutamine are important energy sources for osteoblasts. Oxalate, an LDH inhibitor, shifts the biological energy of bone progenitor cells from glycolysis to oxidative phosphorylation, increasing their capacity for osteogenic differentiation. Overall, it appears that the osteoblast differentiation is controlled by the direction of energy source utilization.

The maturation of osteoblasts is related to bioenergetic metabolism. High transmembrane potential in mitochondria occurs in differentiating osteoblasts. This is supported by the fact that during osteoblast differentiation, mitochondrial oxygen consumption increases [16]. Since there is evidence that the external alternating mechanical stimulation, at the distinct parametric range (around 60 Hz) decreases the maturation of osteoblasts [17], the hypothesis of interrelation between external biophysical stimulation and buildup of mitochondrial transmembrane potential should be further investigated.

ATP content in osteoblasts rises during differentiation, mostly due to aerobic glycolysis. But if a pharmacological suppression of cellular ATP content, with decreased cellular glucose incorporation, is applied by Ro5-4864 (4-chloro derivative of diazepam), cell maturation might still be unaffected [18]. This indicates that other sources of energy, e.g., glycogenic amino acids and fatty acids, might serve as a "second line" protective energetic source. Further investigation of this phenomenon should be of high interest.

The buildup of intracellular biomass, which is supported by macromolecule production and is primarily fueled by glucose, is necessary for cell proliferation. There might be an intriguing hypothesis of the direct relation between mechanical stress and mitochondrial function when fluid shear stress in osteogenic cells generates Ca^{2+}-enhanced mitochondrial Ca^{2+} uptake, which is necessary for the Tricarboxylic acid cycle activity. As a result, the impact of mechanical pressure on osteogenic cells depends on the bioenergetic state of the mitochondria [15].

The osteoblast is an important mesenchymal cell that responds to external biophysical effects. This cell type can be manipulated externally, mainly by mechanical or electromagnetic stimulation [9, 13]. Osteoblasts are responsible for bone matrix elaboration and, therefore, are highly active metabolically (Fig. 1.3). Thus they "orchestrate" bone regeneration and remodeling by interaction with osteoclasts, blood supply, and connective tissue, which comprise the basic multicellular unit (BMU) (Fig. 1.4) [19]. The main role of the osteoblasts to generate bone matrix is induced by external biophysical stimulation. This activity of the osteoblasts is under control of interactions with osteoclasts and osteocytes, which serve as important sensing of the external biophysical stimulation. Therefore, the eventual bone formation is under mutual control of cells of two cells lineages: mesenchymal origin (MSCs which are progenitors of osteoblasts and osteocytes) and hematopoietic origin (macrophages derived from monocytes), and under the overall effect of external biophysical stimulation (Fig. 1.5) [20].

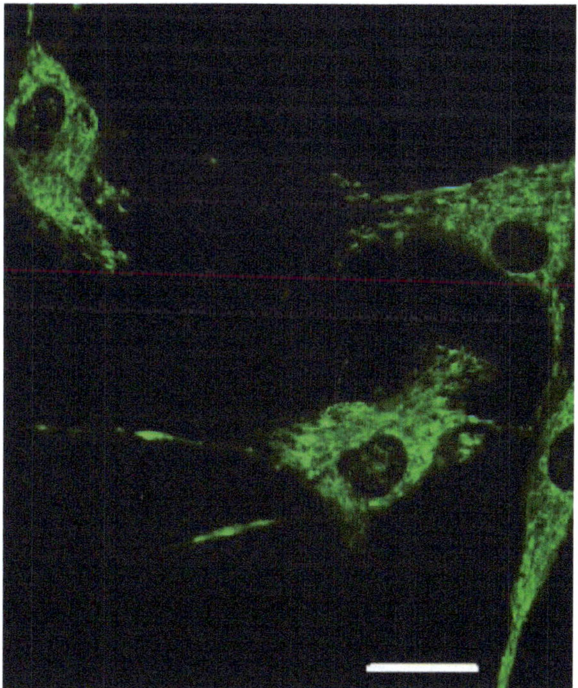

Fig. 1.3 Micrograph (confocal microscopy). Osteoblasts stained for mitochondria by mitotracker green (MTG). High mitochondrial mass is evident, indicating elevated metabolic activity. Scale 20 μm

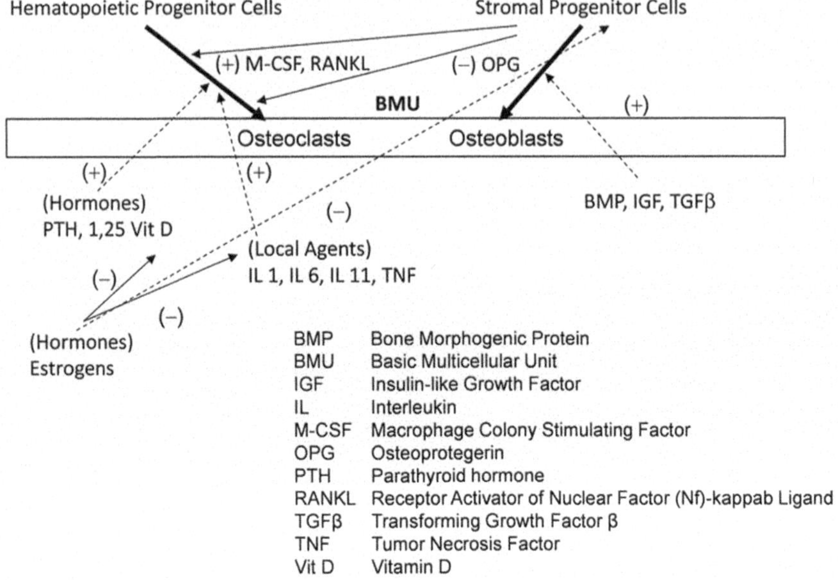

Fig. 1.4 Schematic summary of the interaction of BMU cells [19]. (+): induction, (−): suppression

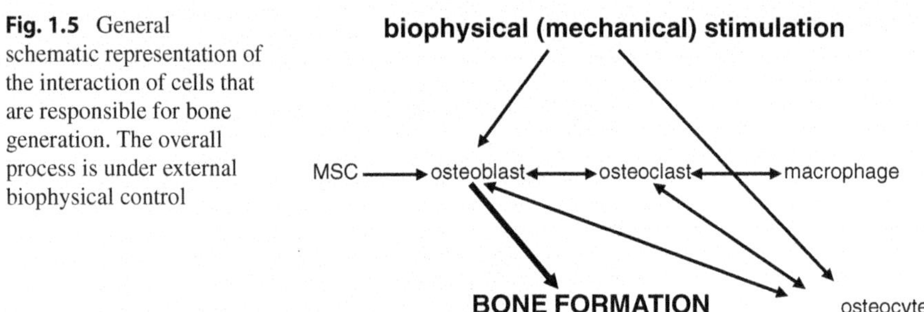

Fig. 1.5 General schematic representation of the interaction of cells that are responsible for bone generation. The overall process is under external biophysical control

In particular, mechanotransduction via membrane PIEZO1 ion channels in osteoblast regulates the expression of collagens II and IX and enhances osteoclast differentiation [6], i.e., mechanical stimulation of osteoblast has a regulative effect on the BMU as a whole. Therefore, osteoblast is an effective candidate for studying the biophysical stimulation of cells with a crucial clinical impact.

Numerous data show that osteoblasts' biophysical stimulation is more effective when applied in a pulsed or alternating high-frequency mode either by electromagnetic field (Table 1.1) or mechanical stimulus (Table 1.2).

The transmembrane flux of electric currents can be initiated by directly applying the electromagnetic field to cells. Electromagnetic stimulation of osteoblast that affects proliferation, maturation, and apoptosis involves two major interconnected pathways. First is the electrocoupling affecting transmembrane ion flux via ion

Table 1.1 Synopsis of selected published data on the response of cultured osteoblasts to alternating electromagnetic field (AEMF) and pulsed electromagnetic field (PEMF) [19]

Refs.	Magnetic flux (mT)	AEMF frequency (Hz)	PEMF frequency: basic/pulses (Hz)	Cellular AP	Cell proliferation
[21]	0.4	–	14.9	↑	↑
[22]	1.8	–	4k/15	↑	↓
[23]	1.8	30	–	↓	Not given
[24]	5	15	–	↑	↓
[25]	5	–	50k/15	↑	↑
[26]	6	–	330/50	Not given	↑

↑ Represents a significant increase; ↓ Represents a significant decrease

channels and membrane receptors, causing membrane depolarization. An additional path is electromagnetic signal propagation to biochemical pathways that activate mitogen-activated protein kinase (MAPK) with subsequential activation of transcriptional factors such as extracellular signal-regulated kinase (ERK) and JNK [33].

Even extremely low alternating electromagnetic flux of 0.2 mT (10 kHz frequency) triggers osteoblast maturation (increase in cellular *alkaline phosphatase*-specific activity) [34].

It is important to emphasize that both mechanical and electromagnetic, alternating biophysical stimulation of osteoblasts are propagated by the canonical WNT/β--catenin signaling pathway leading to osteogenic cellular activity [35, 36]. The WNT/β-catenin signaling pathway is induced by the G protein signaling proteins activated by the extracellular factors, biophysical and humoral (PTH, TSH), across the cellular membrane via the transmembrane domain receptors. This process is related to Ca^{++} flux into cells [37].

Moreover, electromagnetic energy can also be applied to cells by visible light irradiance. Several theories exist regarding the mechanism of cellular photobiomodulation. Although it has been shown that visible light is effective in cellular activation in general, little has been studied regarding the effect of light on cells in the osteoblast lineage. It is apparent from different studies that blue light (453 nm) is effective in enhancing the differentiation of mesenchymal stem cells into osteoblasts through flavinprotein activation [38], while the osteogenic effect of red light (630 nm) involves cytochrome C oxidase in the mitochondria [39].

Thus, according to the facts:

- Cells of mesenchymal origin, in general, can be manipulated biophysically.
- Osteoblasts are highly susceptible mesenchymal cells that respond to alternating biophysical external manipulation.
- Specific optimal biophysical parameters for osteoblast manipulation are not known.

Therefore, the research covering the effects of biophysical stimulation of osteoblasts is of high importance because of its potential clinical impact on governing bone tissue regeneration in pathological conditions when irreparable gaps (critical

Table 1.2 Parametric data of the mechanical induction of proliferation of osteoblast-like cells in vitro in different studies

Refs.	Frequency Hz	Displacement mm	Acceleration mm/s^2	Time hours	Waveshape	Cells	Origin	Parameters investigated
[27]	0.05	0.009[a]	0.00089[b]	48	Square	Osteoblasts	Chicks	DNA, Cell No.
[28]	20–2k	NI	50000	0.12	Sine	MC3T3E1	Mouse	mRNA:c-fos, c-myc
[29]	1	0.0003[a]	0.0112[b]	1.2	NI	Osteoblasts	Human	Cell No.
[30]	1	0.0003[a]	0.0112[b]	0.07	NI	Osteoblasts	Bullock	Cell No.
[31]	0.5	0.003[a]	0.03[b]	6	Square	MC3T3E1	Mouse	DNA

Reproduced with permission from [32]
NI not indicated
[a]Average displacement estimated from strains applied to cells, assuming that the diameters of osteoblasts are in the range of 20–40 μm
[b]Calculated with approximation to sine-shaped vibration

gaps) in the bone should be treated by bone grafting. To promote the basic knowledge of the clinical application of the biophysical methods of osteoblast manipulation for bone regeneration, two essential areas should be addressed:

1. Identification of the specific alternating biophysical parameters for the induction of the phenotypic cell function of human osteoblasts.
2. By using the optimal biophysical parameters for osteoblast stimulation, the authentication of methods for generating viable bone tissue in vitro should provide a safe and effective source for use as an autologous bone graft in vivo.

References

1. Yost MG, Liburdy RP. Time-varying and static magnetic fields act in combination to alter calcium signal transduction in the lymphocyte. FEBS Lett. 1992;296(2):117–22.
2. Liburdy RP. Calcium signaling in lymphocytes and ELF fields evidence for an electric field metric and a site of interaction involving the calcium ion channel. FEBS Lett. 1992;301(1):53–9.
3. Sundelacruz S, Moody AT, Levin M, Kaplan DL. Membrane potential depolarization alters calcium flux and phosphate signaling during osteogenic differentiation of human mesenchymal stem cells. Bioelectricity. 2020;1(1):56–66.
4. Bullard EC. The secular change in the Earth's magnetic field. Geophys Suppl Month Not Royal Astronom Soc. 1948;5(7):248–57.
5. Corrigan MA, Johnson GP, Stavenschi E, Riffault M, Labour MN, Hoey DA. TRPV4-mediates oscillatory fluid shear mechanotransduction in mesenchymal stem cells in part via the primary cilium. Sci Rep. 2018;8:3824.
6. Wang L, You X, Lotinun S, Zhang L, Wu N, Zou W. Mechanical sensing protein PIEZO1 regulates bone homeostasis via osteoblast-osteoclast crosstalk. Nat Commun. 2020;11:282.
7. Carter A, Popowski K, Cheng K, Greenbaum A, Ligler FS, Moatti A. Enhancement of bone regeneration through the converse piezoelectric effect, a novel approach for applying mechanical stimulation bioelectricity. https://doi.org/10.1089/bioe.2021.0019
8. Miroshnikova YA, Nava MM, Wickström SA. Emerging roles of mechanical forces in chromatin regulation. J Cell Sci. 2017;130(14):2243–50.
9. Berkovich Y, Shapira J, Rosenberg N. Chapter 5. Mechanotransduction in osteoblasts. In: Rosenberg N, editor. Experimental methods for mesenchymal cell activation by biomechanical stimulation. Bentham Science; 2016. p. 49–57.
10. Khilan AA, Al Maslamani NA, Horn HF. Cell stretchers and the LINC complex in mechanotransduction. Arch Biochem Biophys. 2021;702:1–7.
11. Gao H, Zhai M, Wang P, Zhang X, Cai J, Chen X, Shen G, Luo E, Jing D. Low-level mechanical vibration enhances osteoblastogenesis via a canonical Wnt signaling-associated mechanism. Mol Med Rep. 2017;16(1):317–24.
12. Luo Y, Zhang Y, Miao G, Zhang Y, Liu Y, Huang Y. Runx1 regulates osteogenic differentiation of BMSCs by inhibiting adipogenesis through Wnt/β-catenin pathway. Arch Oral Biol. 2019;97:176–84.
13. Delaine-Smith RM, Javaheri B, Edwards JH, Rumney VM, RMH. Preclinical models for *in vitro* mechanical loading of bone derived cells. Bonekey Rep. 2015;4:728.
14. Ye J, Xiao J, Wang J, Ma Y, Zhang Y, Zhang Q. The interaction between intracellular energy metabolism and signaling pathways during osteogenesis. Front Mol Biosci. 2022;8:807487. www.frontiersin.org
15. Sautchuk R Jr, Eliseev RA. Cell energy metabolism and bone formation. Bone Rep. 2022;16:101594. https://doi.org/10.1016/j.bonr.2022.101594.

16. Shen L, Guoli H, Karner CM. Bioenergetic metabolism in osteoblast differentiation. Curr Osteoporos Rep. 2022;20:53–64.
17. Rosenberg N, Rosenberg O, Halevi Politch J, Abramovich H. Optimal parameters for the enhancement of human osteoblast-like cell proliferation in vitro via shear stress induced by high-frequency mechanical vibration. Iberoam J Med. 2021;3(3):204–11.
18. Rosenberg N, Rosenberg O, Weizman A, Veenman L, Gavish M. In vitro effects of the specific mitochondrial T.S.P.O. ligand Ro5 4864 in cultured human osteoblasts. Exp Clin Endocrinol Diabetes. 2018;126(2):77–84.
19. Rosenberg N, Rosenberg O, Soudry M. Osteoblasts in bone physiology—mini review. Rambam Maimonides Med J. 2012;3:1–7.
20. Bolamperti S, Villa I, Rubinacci A. Bone remodeling: an operational process ensuring survival and bone mechanical competence. Bone Res. 2022;10:48.
21. Barnaba S, Papalia R, Ruzzini L, Sgambato A, Maffulli N, Denaro V. Effect of pulsed electromagnetic fields on human osteoblast cultures. Physiother Res Int. 2013;18(2):109–14.
22. Lohmann CH, Schwartz Z, Liu Y, Guerkov H, Dean DD, Simon B, Boyan BD. Pulsed electromagnetic field stimulation of MG63 osteoblast-like cells affects differentiation and local factor production. J Orthop Res. 2000;18(4):637–46.
23. McLeod KJ, Collazo L. Suppression of a differentiation response in MC-3T3-E1 osteoblast-like cells by sustained, low-level, 30 Hz magnetic-field exposure. Radiat Res. 2000;153:706–14.
24. Zhang X, Zhang J, Qu X, Wen J. Effects of different extremely low frequency electromagnetic fields on osteoblasts. Electromagn Biol Med. 2007;26(3):167–77.
25. Tong J, Sun L, Zhu B, Fan Y, Ma X, Yu L, Zhang J. Pulsed electromagnetic fields promote the proliferation and differentiation of osteoblasts by reinforcing intracellular calcium transients. Bioelectromagnetics. 2017;38(7):541–9.
26. Suryani L, Too JH, Hassanbhai AM, Wen F, Lin DJ, Yu N, Teoh SH. Effects of electromagnetic field on proliferation, differentiation, and mineralization of MC3T3 cells. Tissue Eng Part C Methods. 2019;25(2):114–25.
27. Buckley MJ, Banes AJ, Levin LG, Sumpio BE, Sato M, Jordan R, Gilbert J, Link GW, Tay TS, R. Osteoblasts increase their rate of division and align in response to cyclic, mechanical tension *in vitro*. Bone Miner. 1988;4:225–36.
28. Tjandrawinata RR, Vincent VL, Hughes-Fulford M. Vibrational force alters mRNA expression in osteoblasts. FASEB. 1997;11:493–7.
29. Neidlinger-Wilke C, Wilke HJ, Claes L. Cyclic stretching of human osteoblasts affects proliferation and metabolism: a new experimental method and its application. J Orthop Res. 1994;12:70–8.
30. Jones DB, Nolte H, Scholubbers JG, Turner E, Veltel D. Biochemical signal transduction of mechanical strain in osteoblast-like cells. Biomaterials. 1991;12:101–10.
31. Stanford CM, Morcuende JA, Brand RA. Proliferative and phenotypic responses of bone-like cells to mechanical deformation. J Orthop Res. 1995;13:664–70.
32. Rosenberg N, Levy M, Francis M. Experimental model for stimulation of cultured human osteoblast-like cells by high frequency vibration. Cytotechnology. 2002;39:125–30.
33. Leppik L, Costa Oliveira KM, Bhavsar MB, Barker HJ. Electrical stimulation in bone tissue engineering treatments. Eur J Trauma Emerg Surg. 2020;46(2):231–44.
34. Rosenberg N. Triggering cultured human osteoblast-like cells' maturation by an extremely low magnitude alternating electromagnetic field. Iberoam J Med. 2020; https://doi.org/10.5281/zenodo.4319702.
35. Lin C-C, Lin R-W, Chang C-W, Wang G-J, Lai K-A. Single-pulsed electromagnetic field therapy increases osteogenic differentiation through Wnt signaling pathway and sclerostin downregulation. Bioelectromagnetics. 2015;36(7):494–505.
36. Duan P, Bonewald LF. The role of the wnt/β-catenin signaling pathway in formation and maintenance of bone and teeth. Int J Biochem Cell Biol. 2016;77(Pt A):23–9.
37. Keinan D, Yang S, Cohen RE, Yuan X, Liu T, Li Y-P. Role of regulator of G protein signaling proteins in bone. Front Biosci (Landmark Ed). 2014;19:634–48.
38. Lewis JB, Wataha JC, Messer RLW, Caughman GB, YamamotoT HSD. Blue light differentially alters cellular redox properties. J Biomed Mater Res B Appl Biomater. 2005;72(2):223–9.
39. Karu TI, Pyatibrat LV, Afanasyeva NI. A novel mitochondrial signaling pathway activated by visible to near-infrared radiation. Photochem Photobiol. 2004;80(2):366–72.

Selected Research Methodologies of Biophysical Stimulation of Osteoblast

2

The aim is to investigate cell maturation, synthetic activity, and cell death rate following the exposure of human osteoblasts to alternating biophysical simulation methods. Mechanical vibration, alternating and pulsed electromagnetic fields, and pulsed photobiomodulation can be used for such purposes. Several versatile standard experimental concepts have been described previously and are used primarily in this type of research.

Using these experimental platforms, with specific biophysical means, viable bone tissue might be generated in vitro and tested in vivo.

The recent reports reflect the stages of the ongoing research that gradually provided important initial insights for determining standardized methods and uniform experimental platforms that could be used for the biophysical manipulation of human osteoblasts by the methods of one tissue engineering.

There are several basic similar standard principles of this series of research projects:

1. The experiments are carried out on monolayer explant cultures of human osteoblast-like cells after the osteoblast characteristics were validated [1].
2. The most common end parameters investigated were cell proliferation, cell death, and cell phenotypic cell function, which were revalidated for the osteoblasts [2].
3. In all the projects, specially designed sources of external application of mechanical, electromagnetic, and light energy are used when the devices are tunable, and all the biophysical parameters are recordable.

N. Rosenberg, *Biophysical Osteoblast Stimulation for Bone Grafting and Regeneration*, https://doi.org/10.1007/978-3-031-06920-8_2

2.1 Osteoblast Explant Cultures

Monolayer culture of adherent to surface osteoblasts is a reproducible method for cell stimulation research. Primary cultures of human osteoblast-like cells originating from chips of cancellous bone collected during surgical interventions are the most common source for human-derived osteoblast culturing (Fig. 2.1). To prevent local metaplastic effects, care should be taken to avoid the proximity of the samples from osteoarthritic joint surfaces in the subchondral area. Usually disposable during surgery, bone samples, each 2–3 g total, are incubated in an osteogenic medium [3, 4] for 20–30 days. Human osteoblast-like cells grow from the chips as a primary cell culture adherent to the plastic tissue culture plates (non-pyrogenic polystyrene) [5]. The polystyrene stiffness in the temperature environment below 100 °C is 35–55 MPa [5], optimal for the osteoblastogenesis range of 25–40 MPa [5].

The explant culture of osteoblasts originates from MSCs that line and/or cover "supporting niches" for hematopoietic stem cells in the bone trabeculae embedded in the bone marrow and, when exposed to osteogenic conditions (osteogenic media, stiff surface, and mechanical stimulation), become committed to the osteoblast lineage [6].

The human bone cell cultures obtained by this standard method express osteoblast-like characteristics, i.e., polygonal multipolar morphology; expression of the enzyme alkaline phosphatase; synthesis of a collagen-rich extracellular matrix with predominantly type I collagen and also small amounts of collagen types III and V; as well as non-collagenous proteins such as sialoprotein (BSP) and osteocalcin [3, 4]. Additionally, these cells demonstrate matrix mineralization in vitro and bone formation in vivo. The osteoblast characteristics of these cells are supported by a positive Von Kossa staining (Fig. 2.2), synthesis of osteopontin (Fig. 2.3), osteocalcin (Fig. 2.4), cellular alkaline phosphatase activity, characteristic multipolar morphology, and adherence to a plastic surface [1].

The cells migrate from the bone chips into the medium and proliferate in culture flasks for 21 days (explant cultures). Then the cells may be placed into multiple well plates, where each well is seeded with 10^3–10^4 cells. This allows the generation of multiple identical samples of cells to be exposed to different experimental

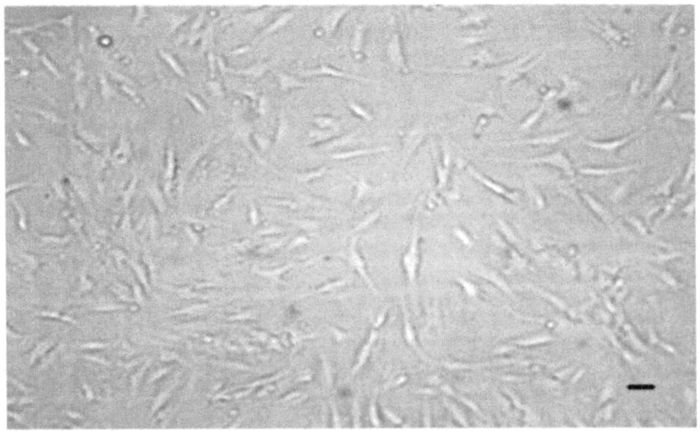

Fig. 2.1 Cell culture's representative microscopic image (scale 20 μm). The characteristic multipolar morphology of the osteoblasts is evident

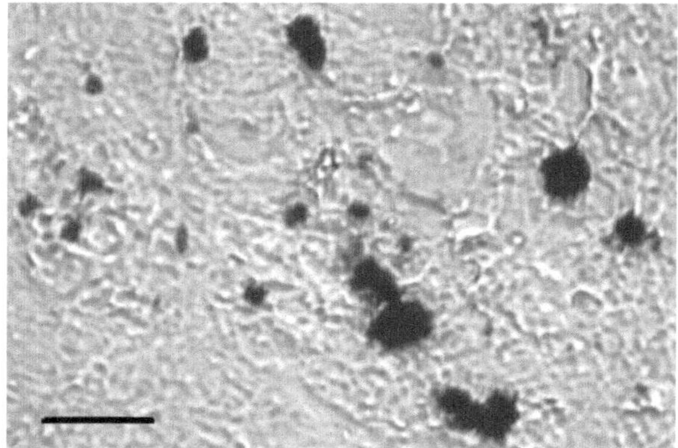

Fig. 2.2 Von Kossa staining of the human osteoblast-like cells in culture. A low-power micrograph shows numerous mineral nodules stained black by silver nitrate (scale 25 μm) [1] (figure source—PhD Thesis authored by Rosenberg Nahum, "High-frequency alternating biophysical stimulation of human osteoblast," School of Pharmacy and Biomedical Sciences, Faculty of Science, University of Portsmouth, October 2021)

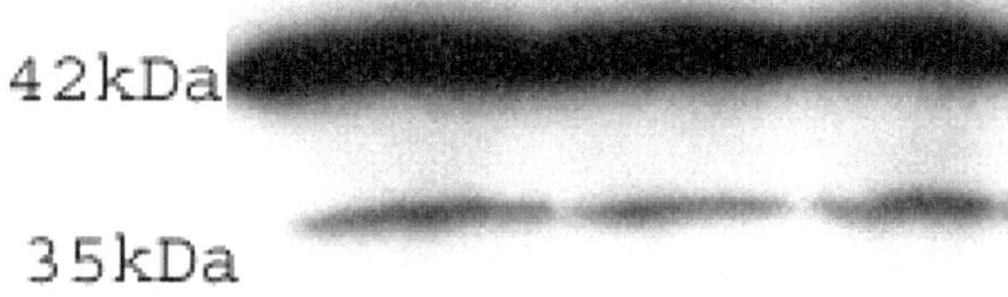

Fig. 2.3 Western blot analysis of osteopontin (35 kDa) expression in the primary cell culture cells. An example of SDS-PAGE of the three samples studied shows osteopontin expression in all samples and representation of expressed β actin (42 kDa) [1] (figure source—PhD Thesis authored by Rosenberg Nahum, "High-frequency alternating biophysical stimulation of human osteoblast," School of Pharmacy and Biomedical Sciences, Faculty of Science, University of Portsmouth, October 2021)

conditions. To avoid cellular dedifferentiation, preferably, the second passage of cells should be used in the experiments. By this method, identical duration, and timing of exposure to experimental conditions in all experiments in a large number of cultured replicates may be implemented. The doubling time of osteoblasts is dependent on seeding density. When seeded up to the 10^4 cells/cm^2 density, the doubling time is around 10 h [7, 8]. Therefore, 4 days of the experiment (96 h) represent ten doubling periods. At this point, the cell culture is close to 2D confluency, the proliferation rate ceases, and the cells start with matrix elaboration. Therefore, the cells reach the maximal proliferation rate in 4 days and activate the phenotypic cell function. Accordingly, the evaluation of the biophysical parameters that affect osteoblast proliferation, cell death, maturation, and phenotypic cell function of cell culture growth at this point might be of high efficacy for drawing meaningful conclusions. Besides bone marrow, there are potential sources for osteoblast monolayer cultures.

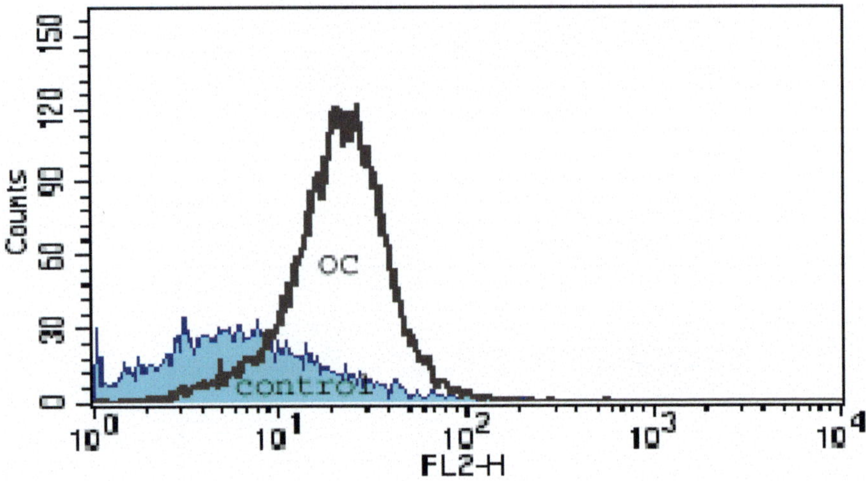

Fig. 2.4 Cytometric analysis of cells positively stained for osteocalcin (oc): 70–80% of the cells were positively stained for oc [1] (figure source—PhD Thesis authored by Rosenberg Nahum, "High-frequency alternating biophysical stimulation of human osteoblast," School of Pharmacy and Biomedical Sciences, Faculty of Science, University of Portsmouth, October 2021). *contr* control non-stained cell

Adipocytes evolve from the same MSCs origin as osteoblasts, following the suppression of the osteogenic signaling of WNT/β-catenin pathway, BMPs, etc., and with parallel induction of adipogenesis by peroxisome proliferator-activated receptors (PPAR) and other transcription factors [9]. Therefore, there is plasticity between osteogenic and adipogenic cell lineages, either direct or through dedifferentiation to fibroblasts. There are three main agents that are capable of transforming mature adipocytes to osteoblasts, i.e., all-trans retinoic acid, BMP 9, and VEGF [10].

An even more appealing source for osteoblast culture is the cells of osteoblast lineage existing in the peripheral blood [11] because of the easy availability of blood samples. The osteoblasts grow as a monolayer primary culture, adherent to a solid plastic surface, from the mononuclear fraction of peripheral blood samples when exposed to the osteogenic media, similar to the described above explant cultures originating from bone marrow samples [12].

Although the biological samples for the latter two methods are easier to harvest, bone marrow-derived osteoblasts are currently more common in research and attempts for clinical use. This is probably expected to change following improvements in osteoblasts culturing techniques from adipose tissue and peripheral venous blood samples.

2.2 End Parameters

There are standard basic parameters that should be investigated in osteoblast response to biophysical stimulation research. These parameters reflect the cell cycle, proliferation rate, maturation, and cell deaths and subsequentially give the basic profile of cellular response to the external biophysical environment.

The equipment for this basic series of tests is readily available, and the methods are highly reproducible.

2.3 Cell Death and Proliferation Estimation

The standard methods for quantitative cell proliferation and cytotoxicity measurements are based on colorimetric measurements of the reduction of a yellow tetrazolium salt to formazan [13]. Alternatively, the total DNA content in each replicate culture can be measured by the fluorescence enhancement method with Hoechst dyes [14], but this method is less accurate because it does not take into consideration the ploidy level of cells during the cell cycle. These are relatively precise methods for the endpoint measurements but cannot be used for the follow-up in the live cultures.

The indirect estimation of cell proliferation in the monolayer live cultures may be done by the qualitative method that considers the number of cells in the culture replica and the rate of cell death.

The number of cells in each culture sample can be estimated by direct counting either by cytometry or by image processing software. The mean value of cell numbers from different microscopic fields in each culture sample after exposure to biophysical stimulation and control samples are recorded and compared. The cell death crude estimation can be done by detecting the Lactate dehydrogenase (LDH) specific activity in the culture media, a marker of cell death due to LDH leakage via damaged cell membranes. The measurements are performed using 340 nm wavelength spectrophotometry of the reduced nicotinamide adenine dinucleotide (NAD), directly proportional to LDH activity [15]. The LDH measurements were validated separately and ensured that this method is suitable for cell death research in osteoblasts (Fig. 2.5) [2].

Thus, the estimation of cell proliferation differences results from the deduction of the change in cell death from the change in the total number of cells in the culture samples. It is important to emphasize that this method is more implacable in the not confluent cultures in monolayer because the individual cells can be distinguished easily by image processing, without cell staining that might interfere with cell viability and metabolism.

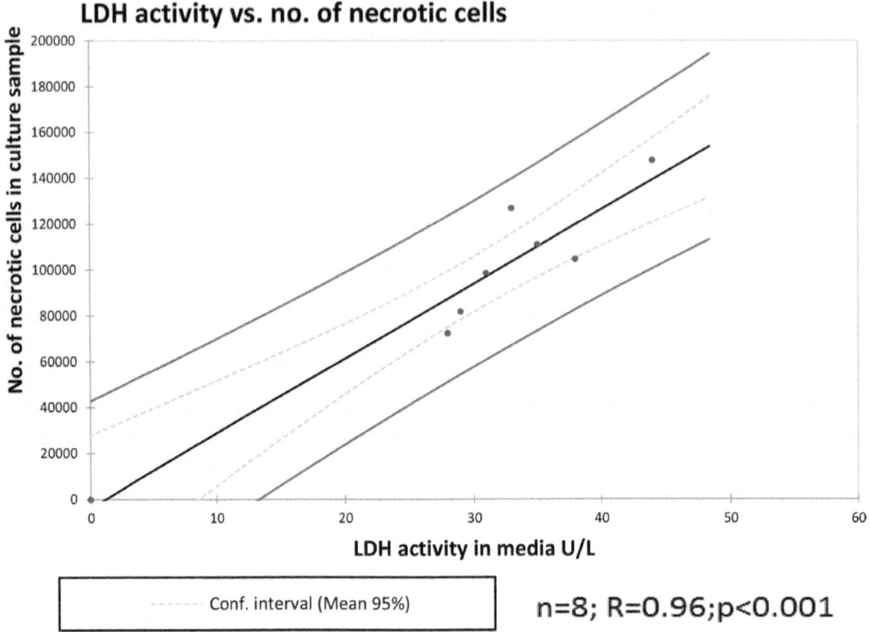

Fig. 2.5 Significant and high correlation was found between LDH activity in culture media and the numbers of non-viable osteoblast-like cells [2]. Reproduced with permission from (© IOSR Journals)

2.4 Cellular Maturation Estimation

Cellular alkaline phosphatase (ALP) specific activity, a marker of osteoblast maturation from progenitors, increases in Stage 2 of the osteoblast cell cycle after Stage 1 cell proliferation ceases [16]. The cellular alkaline phosphatase activity is measured by 410 nm wavelength spectrophotometry [17] or immunohistochemical methods. ALP is a reliable marker for the early stages of osteoblast maturation.

Osteocalcin is an additional specific early marker of osteoblast maturation [18]. Osteocalcin is a 5.8-kDa non-collagenous bone matrix protein secreted by osteoblasts and regulates matrix mineralization. Enzyme-Linked Immunosorbent Assay (ELISA), among other methods, can measure the cellular content of osteocalcin. Immunohistochemical and cytometric methods are also can be easily implemented for this purpose (Fig. 2.4).

2.5 Cell Cycle Profile

Flow cytometric analysis can be used to determine the cell cycle of the osteoblasts [19]. The primary method utilizes a Fluorescence-activated cell sorting (FACS) flow cytometer. Ten thousand cells labeled by propidium iodide are collected per assay point. Then the data is analyzed by computerized means to determine the number of cells in each cell cycle step (Fig. 2.6).

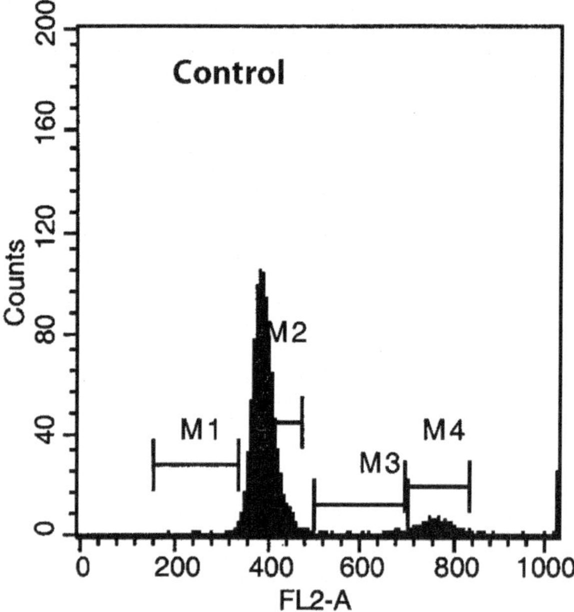

Fig. 2.6 Cell cycle analysis of cultured human osteoblasts. Example histograms of cell cycle phases: M1—pre G1, M2—G1, M3—S, M4—G2/M

All the described above techniques are for the initial "screening" purposes for the identification of the cellular response to external biophysical stimulation and should be further followed by the evaluation of the specific cellular pathways of interest on the transcriptional level.

Numerous related genetic and mitochondrial markers should be used according to the specific research hypotheses, but only after the cellular effect has been crudely identified, i.e., change in cell maturation, proliferation, and cell death rates. For this purpose, those mentioned above can give a relatively simple starting indication.

When proceeding for further determinations of osteoblasts characteristics, a higher level of confidence in the osteoblastic properties of the cells should be based on the determination of cellular Runx2 and Osx transcription factors by using PCR techniques. Furthermore, the MSCs directed to osteogenesis vs. to adipocytogenesis could be detected in the studied culture samples by the cluster of differentiation (CD) markers, such as CD10, CD92, CD31, CD 62E, and CD11B, by flow cytometry [20, 21]. Additionally, other cell type contents in the investigated culture samples should be identified by measuring SOX9 expression, collagen II content by PCR and immunohistochemical analysis, and the Alcian blue histological staining for detection of chondroblastic differentiation [22].

2.5.1 The Assessment of Glycolytic Activity

As mentioned above, the special ability of osteoblasts to utilize glucose as a primary energetic source is almost unique to these cells. This ability of normal, not malignant cells, to use aerobic glycolysis [23] enables to employ techniques that measure glucose cell incorporation for the estimation of cellular metabolic activity. For this purpose, the method of measurement of [^{18}F]-fluorodeoxyglucose incorporation into human osteoblast is a powerful and relatively simple experimental tool [24].

Several processes, including diffusion, active transport, and a quick, facilitated transfer through the family of plasma membrane glucose transporters (GLUT), can be involved to incorporate glucose into cells [25].

One of the four forms of the enzyme hexokinase, each of which is specific to a different cellular type, turns the intracellular glucose into glucose-6-phosphate. Utilizing ADP, which is pumped from the mitochondria through the voltage-dependent anion channel (VDAC), under the control of insulin or insulin-like growth factors, this energetically irreversible process occurs on the mitochondrial membrane [26].

Glycolysis and the pentose phosphate shunt subsequently make the generated glycose-6-phosphate accessible for further metabolism. Because the irreversible rate-limiting stage of the incorporation of glucose into the glycolysis pathway is the hexokinase-mediated conversion of glucose into glycose-6-phosphate, it may be a good indicator of the rate of incorporation of glucose into the cell for estimating its energy consumption. [^{18}F]-fluoro-2-deoxy-D-glucose (FDG), which is a positron emitter tagged glucose analog, enters a live cell through the same enzymatic and transport pathways as glucose and is trapped intracellularly [27]. Therefore, the content of the trapped in cells FDG can be measured by a γ counter. The percentage of the initial dose of FDG in a culture retained in cells indicates cellular glucose incorporation. When cellular FDG incorporation is significantly diminished, for example, by the treatment with protoporphyrine IX 10^{-5} M, which is known to have an apoptotic effect by changing mitochondrial membrane permeability [28], the parallel decrease of ATP content in the osteoblasts is detected [29] (Fig. 2.7).

Therefore, the use of FDG incorporation studies in osteoblasts is a relatively easy and highly effective tool for monitoring the osteoblasts' metabolic activity.

2.5.2 Experimental Setups for External Application of Alternating Biophysical Energy

A particular interest for investigating biophysical stimulation of osteoblasts is alternating spectra above the infrasonic range, i.e., 10–60 Hz, which also overlaps the range of the natural frequency of the osteoblasts (21–41 Hz) [30]. This range of applied energy reflects the trophic effect of involuntary contracting frequencies of skeletal muscles at rest on the adjacent bone [31]. Without this effect, as in paralyzed limb, pronounced bone depletion occurs and is reflected as osteoporotic changes. Theoretically, the trophic effect of the contracting muscles can be transferred to bone cells either by the traction force of muscle attached to bone and/or by electromagnetic fields, which are locally generated between the contracting muscle and bone. It is unclear which of these proposed modes of osteoblast stimulation is more important for bone maintenance.

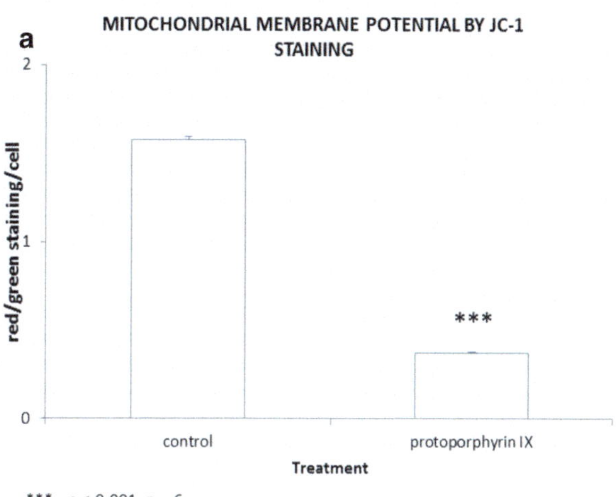

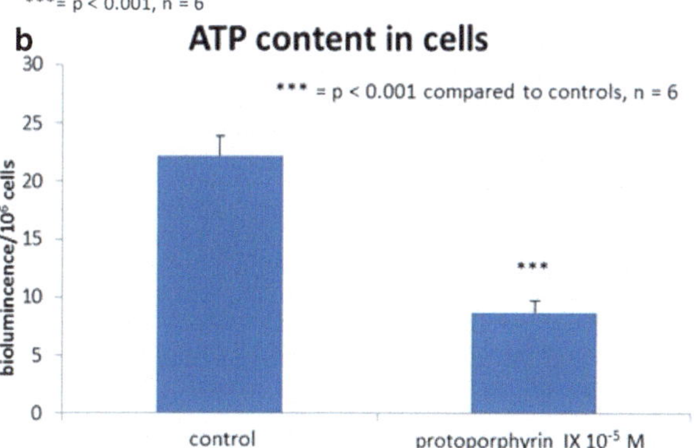

Fig. 2.7 (**a**) Mitochondrial membrane depolarization following exposure to protoporphyrin IX (PPIX) (10^{-5} M) in comparison to vehicle controls. (**b**) ATP content is decreased in cells exposed to PPIX (10^{-5} M) [29]

Accordingly, the research of biophysical stimulation of osteoblasts in this mode of alternating energy environment has the potential for clinical applications, especially in the treatment of osteoporosis and bone fracture healing.

2.6 Visible Light Irradiance

From all the external physical methods of osteoblast stimulation, the exposure of cells to alternating visible light is the easiest to perform but the least investigated. The research attempts to investigate the photobiomodulation in cells of mesenchymal origin utilized different light modes with different irradiance power, e.g., He-Ne laser (632 nm wavelength) with 5.3 mW irradiance [32], visible light with a bandwidth of 400–800 nm at a power of 4.8 J/cm^2 [33]. There is evidence that there is a direct effect of absorption of various wavelengths by cellular chromospheres from these studies.

There is also some evidence that visible light irradiance directs the progenitor cells and mesenchymal stem cells (MSCs) toward osteoblast differentiation.

Moreover, different spectra ranges of visible light have a separate effect on cells with the overall osteogenetic effect, i.e., green light (525 nm) enhances gene expression, blue light (470 nm) enhances osteoblast maturation (calcium deposition and ALP activity) [34].

Overall it is apparent from different studies that blue light spectrum (wavelength of 450–485 nm) is effective in enhancing the differentiation of mesenchymal stem cells into osteoblasts through flavinprotein activation [35], while the osteogenic effect of red light spectrum (wavelength of 625–750 nm) involves cytochrome C oxidase [36].

The mentioned above rationale for the enhanced photobiomodulation by visible light irradiance by a pulsed protocol was clarified in several studies that identified activation of ChR2 channels with a sequence of brief pulses of light irradiance in neurons [37]. This type of photobiomodulation involves the generation of photocurrents into cells [7]. Furthermore, photobiomodulation following pulsed irradiance in the visual spectrum is mediated by the enhanced electron transfer in the redox apparatus of the cytochrome c oxidase in the mitochondria [36].

Although experimental data on photobiomodulation in osteoblasts is scarce, there is a general indication that pulsed light irradiance elicits variable cellular responses, e.g., enhanced cell proliferation and/or differentiation. These effects are separate and distinguishable following irradiance at different frequencies and spectral ranges of the applied visible light [38]. The logical explanation for these phenomena should be related to the ability of cellular photobiomodulation apparatus to absorb the irradiance energy and propagate it into the electrical currents, transmembranous and intracellular. Therefore, in the similarity to cellular mechanotransduction a new term for "cellular phototransduction" might be defined. This interesting hypothesis should be based on solid experimental evidence, which currently does not exist. Naturally, for this purpose, reproducible experimental setups are essential.

In general terms, such a setup should include a stage mounted with transparent to visual light well plates, containing cultured osteoblasts adherent to a solid surface in a monolayer, with the possibility of being exposed to a vertically directed light source in an otherwise dark environment. The light intensity and the distance to the cultures should be tuneable to create a desirable irradiance on the cells. The irradiance energy should be measured from the surface of the culture container. Light filters can be used to investigate the specific spectra of the applied irradiance, and the frequency of pulsed irradiance can be determined by a pulse generator (Fig. 2.8).

By using this relatively simple experimental setup, narrow spectra of low-intensity pulsed (40 Hz) LED (Light Emitting Diode) light irradiance that causes photobiomodulation in the osteoblast were identified, i.e., an increase in the number of cells and cell death by green spectrum range with diffuse transmittance (560–650 nm, maximal cell irradiance 0.4 W/m^2) and a decrease in osteoblast maturation by a blue range of spectrum irradiance (alkaline phosphatase decrease following irradiance with diffuse transmittance 420–580 nm, maximal cell irradiance 0.5 W/m^2) [38].

These results indicate a low-intensity threshold of photobiomodulation of osteoblasts in vitro by 40 Hz pulsed irradiance. Since the membranal stimulation can be

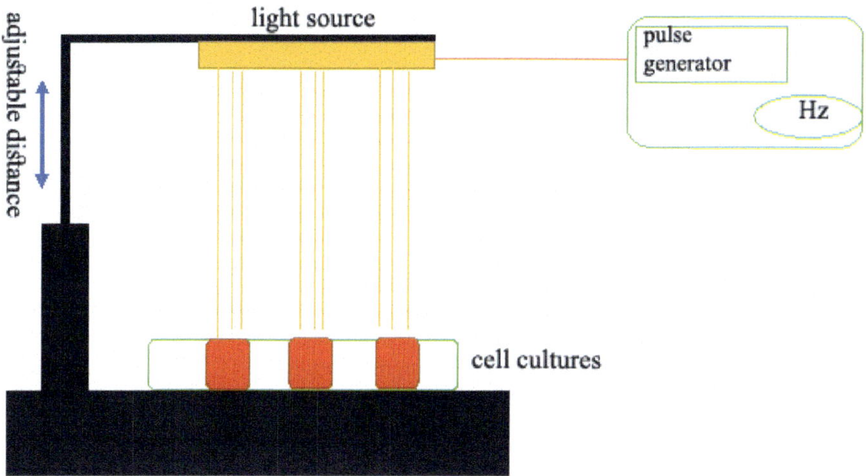

Fig. 2.8 Schematic representation of the experimental setup for photobiomodulation studies on human osteoblasts

produced by white light intensity as low as 0.6 mW/cm^2 in the light-sensitive neural cells [39], it can be assumed that osteoblasts, which are probably less sensitive than visual neurons to light, will react to a higher level of irradiance for the triggering of the photobiomodulation effects.

2.7 Electromagnetic Field

The goal of extensive research is to find the best electromagnetic field (EMF) settings for improving osteoblast maturation, synthetic activity, and proliferation. For this goal, the alternating EMF (AEMF) and pulsed EMF (PEMF) applications are successful (Table 1 in "Historic insights"), but the specific propensity pattern remains unclear.

When administered in an alternating mode of 30 Hz, the same maximum magnetic flux of 1.8 mT produces a decrease in cellular alkaline phosphatase (AP) (the early marker of osteoblast maturation) activity [40]. It has the opposite enhancement effect on cellular AP activity following exposure to PEMF (basic frequency 14 kHz in pulses of 15 Hz) [41].

A higher intensity 5 mT AEMF (15 Hz) and 5 mT PEMF (basic frequency 50 kHz in pulses of 15 Hz) create a similar increased cellular AP activity, but with opposite effects on cell proliferation, i.e., PEMF induces and reduces cell proliferation [42, 43].

Furthermore, the 6 mT PEMF (with greater intensity) had a substantial incremental effect on cell proliferation [9].

The suggested "window effect" theory of the nonlinear nature of modulation EMF frequency and intensity, which is related to the characteristics of calcium transmembrane flux, appears to explain this apparent puzzling lack of clear uniform tendency of cell response to AEMF or PEMF with different frequency and magnetic flux parameters [44].

This phenomenon's specific cellular mechanism is unknown. However, there is important evidence that different ranges and combinations of AEMF and PEMF frequencies and intensities effects varied cellular proliferation, metabolic activity, and cell cycle modulation responses. Recognizing these combinations could lead to the development of effective therapeutic strategies for improving bone regeneration.

In cultured osteoblasts, even a very low 14.9 Hz PEMF intensity of 0.4 mT causes a considerable increase in cell maturation (as measured by cellular AP activity) and proliferation [45]. Because the control cultures are only exposed to a background earth magnetic field of 0.07 mT [46], the intriguing question is if there is an intensity lower than 0.4 EMF but greater than the background EMF that could cause human osteoblasts to enhance proliferation and/or metabolic activity. The answer to this question could help to establish the EMF-triggering values that drive a cellular reaction.

The investigation of the effect of alternating and pulsed electromagnetic fields (AEMF and PEMF, respectively) on cultured osteoblasts also revealed a distinct high-frequency (20–30 Hz) range that reduces cell death and cellular maturation and shifts the cell cycle toward the G1 Phase (Fig. 2.9) [47].

A specially designed system can be used for electromagnetic field delivery to human osteoblast-like cells in culture by placing a coil under the container with cultured cells adherent to the surface in a monolayer (Fig. 2.10).

By this method, different electromagnetic fields with different profiles, e.g., constant EMF, PEMF, and AEMF, can be applied to the cultured cells, and the magnetic flux can be detected and measured. When PEMF is used, its spectrum frequencies can be differentiated by Fast Fourier Transform (Fig. 2.11).

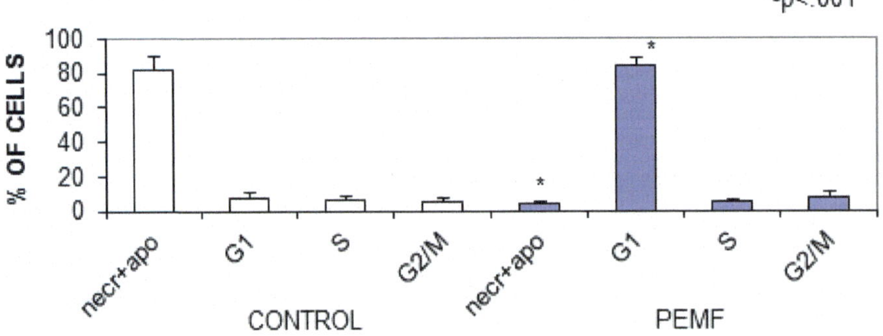

Fig. 2.9 Cytometric profile of cells exposed to PEMF (20–30 Hz) and control culture. Shifting of the cell cycle toward the G1 phase in the cells exposed to the PEMF is evident [47]. *necr* necrotic cells, *apo* apoptotic cells

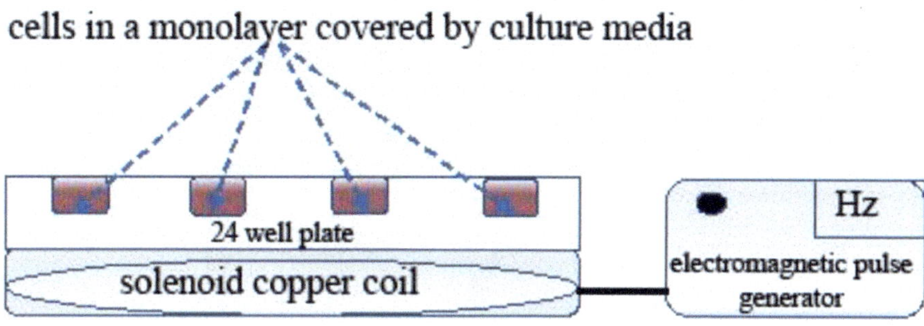

Fig. 2.10 Twenty-four well plate with cultured osteoblast-like cells is mounted on coils that deliver a controlled electromagnetic (EM) field—schematic representation

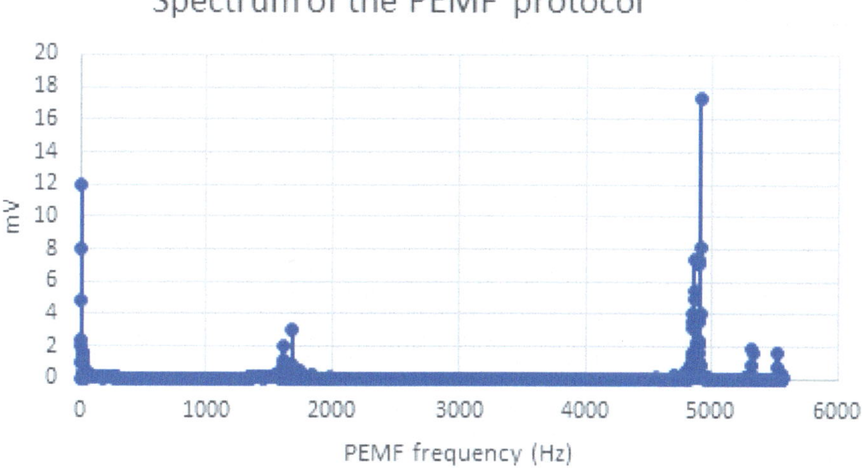

Fig. 2.11 An example of a spectrum of the pulsed electromagnetic field (PEMF) excitation protocol (by Fast Fourier Transform of the EMF recorded profile [48]. *Hz* hertz

When the specific 10–60 Hz range of alternating electromagnetic field effect was investigated in cultured human osteoblasts, by using this type of versatile experimental setup, the triggering for human osteoblast activation for maturation by an extremely low AEMF (10 kHz) is at least 0.2 mT, and for proliferation by (14.9 Hz), PEMF with the maximal intensity of 0.4 mT were detected [47, 48].

2.8 Mechanical Stimulation

An important characteristic that allows mesenchymal cells, especially osteoblasts, to react to controlled mechanical stimulation is their tendency to adhere to solid surfaces and produce monolayer orientation. This ability provides a relatively versatile setup for applying mechanical stimulation, either by controlling the stiffness of the container surface or by applying an external mechanical force to the container, which is transferred from the container to the adherent cells.

The cell–surface interface at the adherence sites is not a solid structure but should be considered a "spring-like" interface (Fig. 2.12); therefore, the external force transferred to the cell via the solid surface is not equivalent but somewhat similar to the external force. The exact mechanical modification of the force characteristics at the surface–cell interface is unknown and should be further investigated. Currently, by simplifying the research protocols, the mechanical force applied to the container

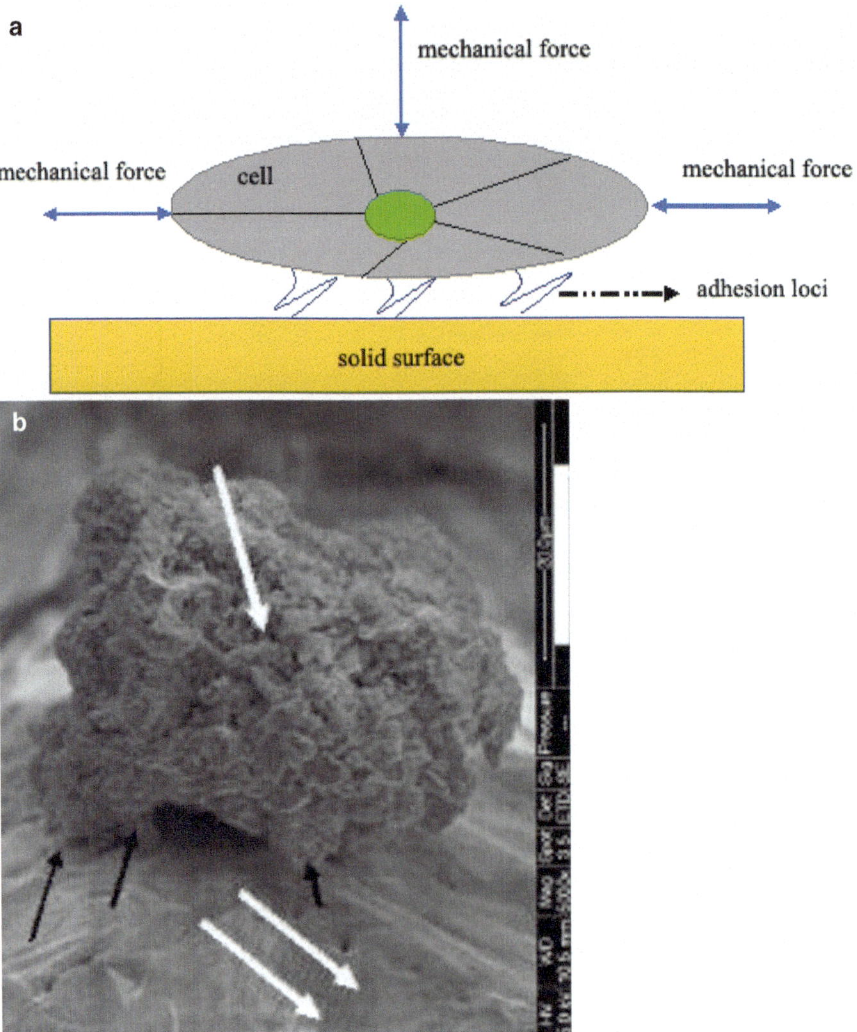

Fig. 2.12 (**a**) Schematic representation of cell exposed to external mechanical shearing forces when adherent to a solid surface. (**b**) Scanning microscopy micrograph of osteoblast (white arrow) attached to a solid surface (two white arrows). The adhesion loci are apparent (black arrows)

surface is considered fully transferred to the adherent cells. The widely used experimental methods for studying the mechanical stimulation of osteoblasts utilize the stretching of cells adherent to elastic membranes (Fig. 2.13) or the implementation of controlled external fluid flow to cell cultures (Fig. 2.14). In both, the cellular activation mechanism is based on shearing forces and subsequential cellular deformation, which causes cytoskeletal activation by transmembrane electrical currents [49, 50]. These experimental methods require relatively sophisticated hardware with essential special handling and complex experimental protocols. But they are usually unable to produce uniform stretching forces on all the cells in culture and are commonly restricted to mechanical loading frequencies of up to 5 Hz. It has been found that the strain values from the actuator input to the stretchable membrane with the adherent cells differ from the actual strain on the cells. Therefore, for the reliable interpretation of the data, a calibration pressure–strain curve should be generated for each specific experiment. Thus, the handling of experiments of osteoblast stimulation by stretchable membrane might provide data that can't be tunable enough and more difficult for compare [51].

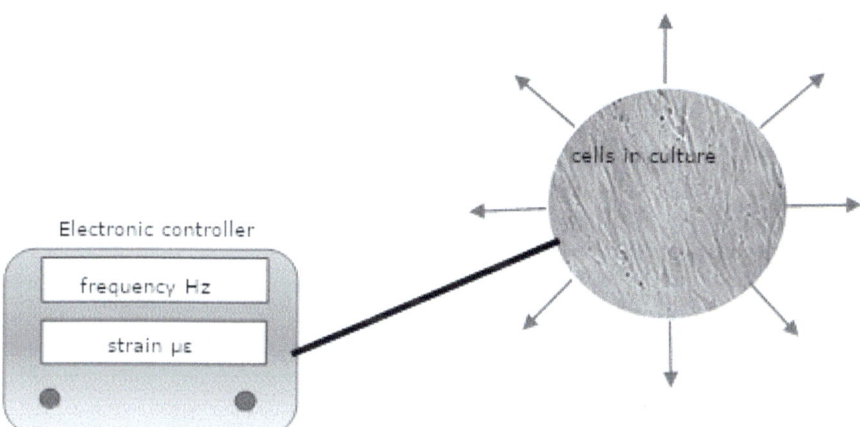

Fig. 2.13 Cells adherent onto stretchable surface exposed to alternating strain in radial direction (arrows)

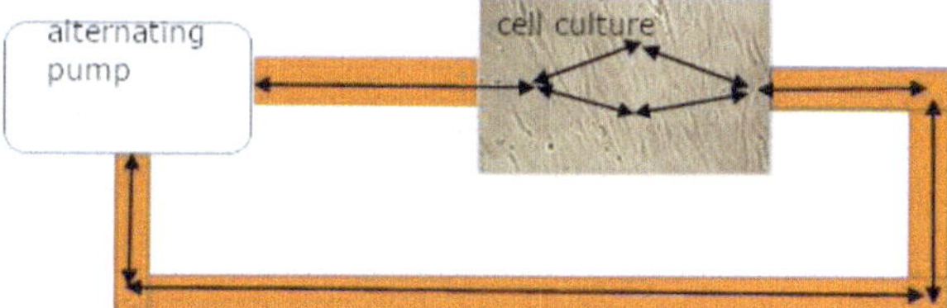

Fig. 2.14 Cells in monolayer adherent to a solid surface are exposed to alternating culture media flow (arrows) in a close loop setup. The alternating fluid flow is controlled by a pump

These methods are insufficient for studying mechanical stimulation in a higher range of frequencies of alternating forces. Another, a more versatile experimental approach that is highly reproducible and has a broader range of mechanical frequencies that could be applied to the cultured osteoblasts in monolayer, is based on applying a controlled vibration force in a defined range of mechanical parameters [52]. In this model, the cellular deformation is caused by shearing forces applied by an extra intracellular fluid flow induced by the one-dimensional accelerated vibration movement of cells adherent to a solid (plastic) surface.

The simplified concept of a cell adherent to a solid surface is as a "lipid bubble" attached to the surface (Fig. 2.15). This is a convenient approach for the theoretical mathematical calculations for estimating strains and deformations that the cell can withstand in physiological conditions. These calculations and experimental observations show that up to 5% strain on a cell is possible without cell membrane rupture [53].

Obviously, because the actual adherence of the osteoblasts to the matrix cannot be defined as solid but as a "spring-like" via adhesion loci (Fig. 2.12), the actual mechanotransduction from the surface to the cytoskeleton is not identical to the grossly measured mechanical parameters of the supporting matrix. The direct measurements of the spectra of cell membrane integrins (adhesion molecules which are part of the adhesion loci) vibration, by using the atomic force microscope readings analysis, might be a promising method for direct measurements of the parameters of mechanotransduction from the matrix to the cytoskeleton [54]. This promising experimental concept should be further clarified in future studies.

Since the cells in this model are under the same environmental conditions, they are exposed to the same magnitude of force during the vibration force application on a culture well plate. Therefore, large numbers of cells can be studied under the same mechanical conditions. In this method, the well plate with cultured cells is mounted on a horizontally oriented actuator. An amplifier controls the vibration movement of the actuator (shaker). A piezoelectric accelerometer and/or

Fig. 2.15 A simplistic presentation of the cell as a lipid bubble (soap) adherent to a moving surface and deforms according to internal and external fluid effects

displacement transducer (Linear Variable Differential Transformer—LVDT). The moving stage with a mounted container for cell culture should move on a low friction interface (Fig. 2.16) [55–57].

In this open-loop setup, not only are the delivered mechanical parameters recordable but the actual data of the supporting surface movement, which reflects the movement of cells, can also be measured online from the reading of the accelerometer or displacement sensor (Fig. 2.17). By using this method, it has been revealed

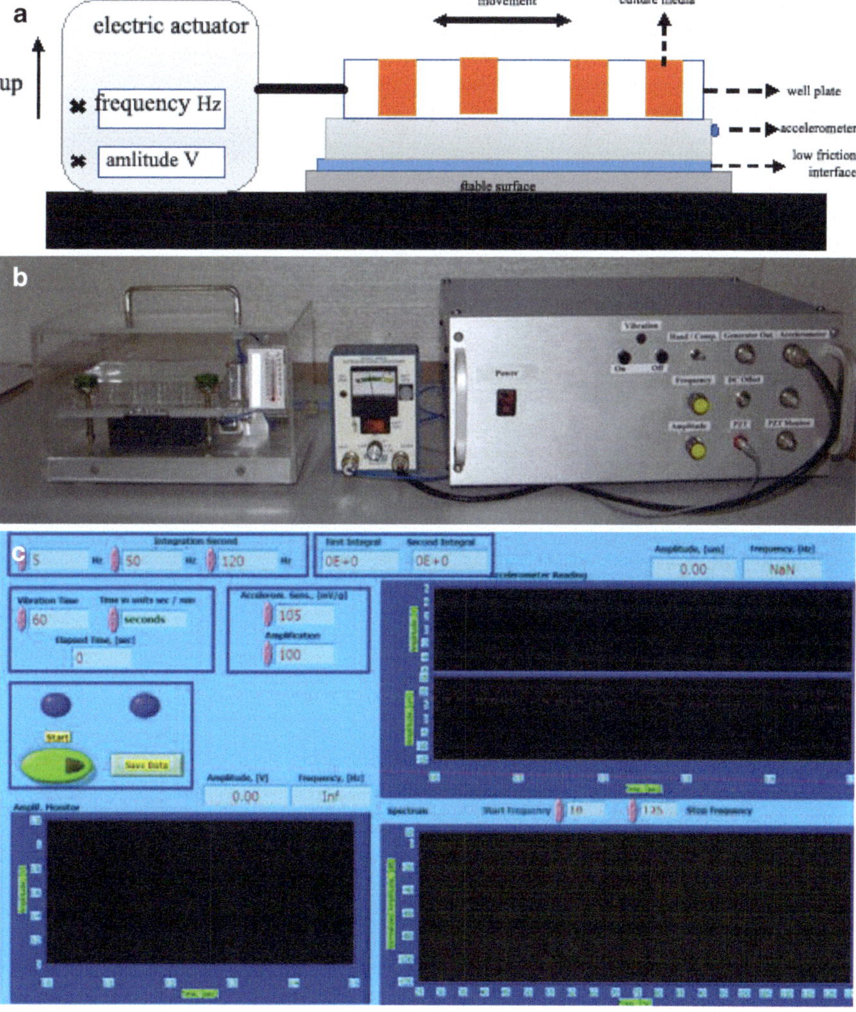

Fig. 2.16 (**a**) Schematic presentation of the experimental setup for the alternating mechanical stimulation (vibration) of cultured osteoblast-like cells. (**b**) Custom-designed experimental device for osteoblast stimulation by vibration according to the principle described in scheme (**a**). (**c**). Control panel for tuning and controlling applied and detected vibration parameters in the device is shown in (**b**)

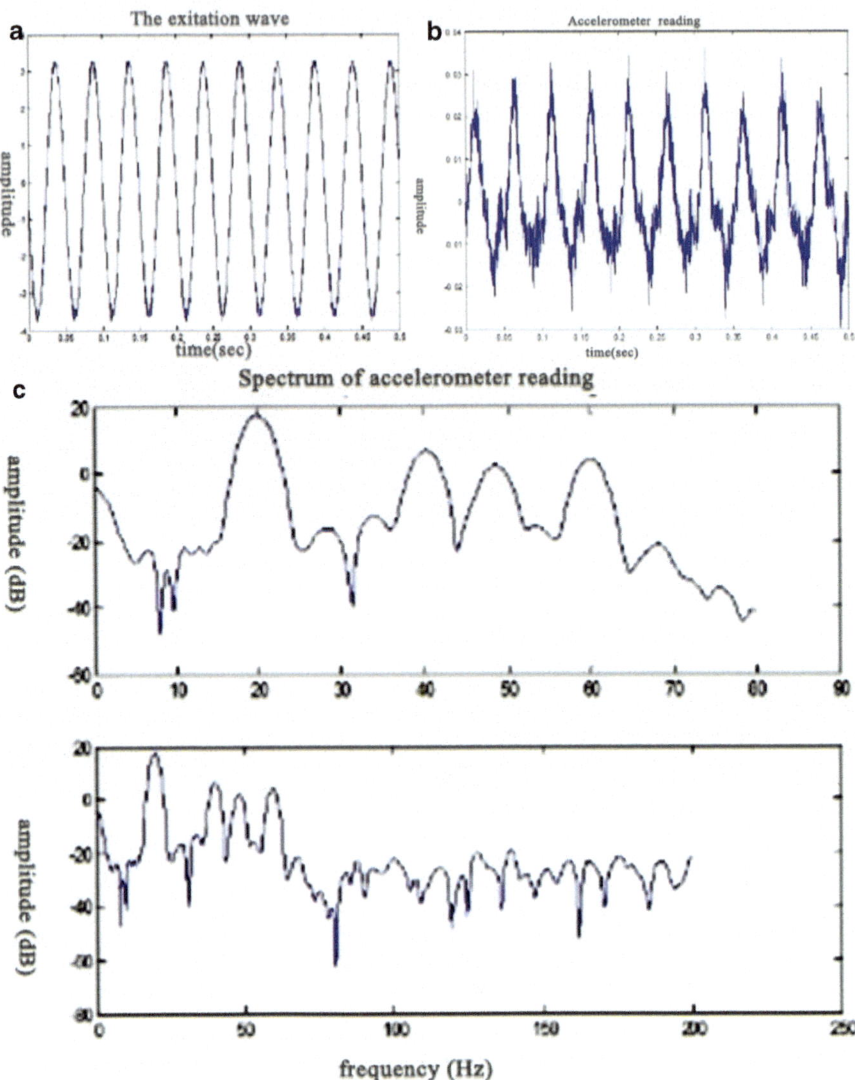

Fig. 2.17 Example of the profile of 20 Hz excitation by the actuator (**a**) and the measurements readings by the accelerometer attached to the moving stage (**b**), which represents the actual mechanical force applied to cells adherent to the moving surface and appears close to the excitation signal from the actuator (amplitude readings represented in $V \times 10^{-1}$). The maximal amplitude of displacement is 25 ± 5 μm. The spectra of the vibration frequencies applied to the cells (**c**) show that 70% of the frequencies, up to 70 Hz frequency, are in the range of 10–30 Hz [55] (figure source—PhD Thesis authored by Rosenberg Nahum, "High-frequency alternating biophysical stimulation of human osteoblast," School of Pharmacy and Biomedical Sciences, Faculty of Science, University of Portsmouth, October 2021)

that external 50–70 Hz alternating mechanical force that causes peak-to-peak acceleration of 0.03 g with a maximal displacement of 10–30 μm enhances osteoblast proliferation with a decrease of cell maturation [55] (Fig. 2.18). This kind of experimental data is significant for the future application of mechanical stimulation of osteoblasts for bone regeneration in clinical use.

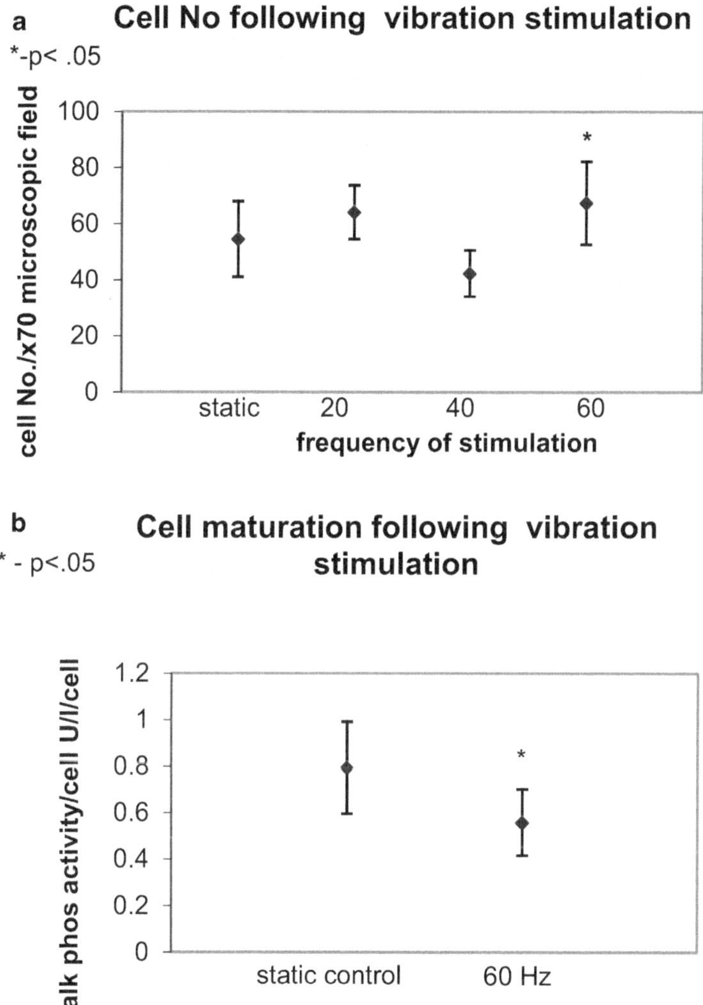

Fig. 2.18 Proliferative (**a**) and maturation (**b**) response of osteoblasts in a monolayer following stimulation by mechanical external vibration. A distinct range of around 60 Hz of vibration causes increased proliferative response with diminished maturation of cells (expressed by the cellular alkaline phosphatase activity). Mean values with standard deviation are presented (n = 12)

2.9 The Strengths and Weaknesses of the Methodologies

The described three methodologies for stimulating cultured osteoblasts by alternating external biophysical energy (electromagnetic, light, and mechanical) allow, by relatively easy-to-construct experimental methods, to expose and study the cellular response in uniform conditions for a large number of cell culture replicas in relatively high, above infrasonic, frequency spectrum. This range of frequencies is like the natural cell frequency and is normally required for the physiology of bone maintenance, as mentioned above. Therefore, the research utilizing the biophysical cell stimulation by alternating 10–70 Hz energy is particularly interesting.

There is already experimental evidence of the compelling mode of pulsed biophysical stimulation in the 20–60 Hz range that could modulate osteoblast phenotypic cell function and proliferation. These reports provide an intriguing foresight regarding the value of the effective biophysical modalities and parameters, the method of their application to cells, and the nature of cytoskeletal involvement in mechanotransduction for future clinical applications.

Currently, the research on the osteoblast response to alternating biophysical stimulation in the range above infrasonic frequencies is in its initial steps, and the precise cellular pathways involved are still elusive [58, 59]. But there are already experimental indications that different sub-ranges of vibration parameters are effective for osteoblast proliferation and phenotypic cell function stimulation, and they do not overlap [57]. The proliferation and phenotypic cell function are found to be stimulated by distinct mechanical stimuli in the 20–70 Hz range.

Additionally, there is evidence that cytoskeletal components propagate external biophysical stimulation by mechanical vibration through the interaction of microfilaments and microtubules [60]. The total DNA content in cultured replicates of osteoblast-like cells following their exposure to enhanced mechanical stimulation by vibration at a 20-Hz frequency (25–30 μm of displacement amplitude), with and without added specific microtubular and microfilament polymerization blockers (colchicine and cytochalasin D, respectively) was compared. The results revealed the essential and unique role of the microtubular component of the cytoskeleton in mechanotransduction for proliferation by showing that colchicine blocked the expected increase in DNA content after mechanical stimulation of the cultured replicates, without altering total DNA content in replicates under static conditions.

Thus, the series of experiments showed that the application of alternating biophysical energy in the frequency range of 20–70 Hz causes proliferative and phenotypic effects in human osteoblasts. The specific optimal biophysical parameters, expressing the amount of energy applied, as represented by the different spectra of the applied vibration, light, and electromagnetic field, were detected. Importantly, all three modalities of the designed experimental setups are easily reproducible for further experiments on large numbers of cultured osteoblasts.

Since the information on the biophysical means to enhance proliferation and phenotypic cell function of these cells has been obtained, it is logical to proceed to research on the ability to stimulate the osteoblasts to generate a bone matrix, with the subsequential aim of generating in vitro bone-like material suitable for implantation in vivo.

Therefore, the subsequent conclusion is that by applying biophysical stimulation to cultured inorganic matrix osteoblasts, as described above, and by using the modified experimental setup, live bone tissue can be generated in vitro [61].

The research projects designed according to the hypothesis of effective osteoblast stimulation by alternating external energy application in the spectra above infrasonic frequencies should be directed to identify the specific biophysical parameters for the implementation of tissue generation in vitro, with the perspective of implantation in vivo. The initial evidence for effectivity of this chain of objectives has already been accomplished. A fundamental issue for identifying the exact cellular pathway was only partly presented in the existing reports, showing the importance of the microtubular component of the cytoskeleton for the propagation of the external biophysical signal (before this research, this ability was attributed mainly to the microfilament component of the cytoskeleton). Obviously, further large-scale studies should be carried out to close the gap of knowledge in this area [58, 59]. By recognizing these findings, applying the optimal biophysical parameters in a clinical setup by a separate finetuning of osteoblast proliferation and maturation might have a therapeutic value in enhancing impaired bone regeneration [62].

The most challenging and not resolved issue in these experimental models is the measuring of the mechanical effect of the moving fluid over cells in the incubation chamber. In osteoblasts, the primary cilium is the sensing cellular interface for this purpose [63]. The cilium is an extracellular microtubular extension from the centrosome acting, among other, as a flow sensor via stretch-activated Ca^{2+} channels. Currently, the difficulty of the measurements of cilia mechanics causes an enigmatic area for estimating the contribution of fluid flow, which might be turbulent when alternating movement is applied to the solid support of the cultured osteoblasts in a monolayer, to cellular deformation and subsequential strain on cells. Currently, the lack of ability to measure the exact contribution of the fluid flow on strain generation is the main obstacle to clarifying the exact mechanical effect on cultured cells subjected to mechanical stimulation by different experimental setups.

References

1. Rosenberg N, Soudry M, Rosenberg O, Blumenfeld I, Blumenfeld Z. The role of activin A in the human osteoblast cell cycle: a preliminary experimental *in vitro* study. Exp Clin Endocrinol Diabetes. 2010;118:708–12.
2. Rosenberg N, Hamoud K, Rosenberg O. Quantitative expression of cell death by LDH activity. IOSR J Pharm Biol Sci. 2016;11:46–8.
3. Gundle R, Stewart K, Screen J, Beresford JN. Isolation and culture of human bone-derived cells. In: Beresford N, Owen ME, editors. Marrow stromal cell culture. Cambridge: Cambridge University Press; 1998. p. 43–66.

4. Yamanouchi K, Satomura K, Gotoh Y, Kitaoka E, Tobiume S, Kume K, Nagayama M. Bone formation by transplanted human osteoblasts cultured within collagen sponge with dexamethasone *in vitro*. J Bone Miner Res. 2001;16:857–67.
5. Properties of Polystyrene (polymerdatabase.com).
6. Wu JY, Scadden DT, Kronenberg HM. Role of the osteoblast lineage in the bone marrow hematopoietic niches. J Bone Miner Res. 2009;24(5):759–64. https://doi.org/10.1359/jbmr.090225.
7. Troh A, Tsai HC, Wang LP, Zhang F, Kressel J, Aravanis A, Santhanam N, Deisseroth K, Konnerth A, Schneider MB. Tracking stem cell differentiation in the setting of automated optogenetic stimulation. Stem Cells. 2011;29(1):78–88.
8. Abdul Wahab R, Rozali NAM, Senafi S, Abidin IZZ, Ariffin ZZ, Ariffin SHZ. Impact of isolation method on doubling time and the quality of chondrocyte and osteoblast differentiated from murine dental pulp stem cells. Peer J. 2017;5:1–17.
9. Suryani L, Too JH, Hassanbhai AM, Wen F, Lin DJ, Yu N, Teoh SH. Effects of electromagnetic field on proliferation, differentiation, and mineralization of MC3T3 cells. Tissue Eng Part C Methods. 2019;25(2):114–25.
10. Lina DPL, Dassa CR. Transdifferentiation of adipocytes to osteoblasts: potential for orthopaedic treatment. J Pharm Pharmacol. 2018;70:307–19.
11. Suriyachand K, Eamwijit T, Paisooksantivatana K, Hongeng S, Bunyaratavej N. A study of correlation of osteoblasts from peripheral blood with related bone turnover markers. J Med Assoc Thail. 2011;94(Suppl 5):S71–5.
12. Wu G, Pan M, Wang X, et al. Osteogenesis of peripheral blood mesenchymal stem cells in self assembling peptide nanofiber for healing critical size calvarial bony defect. Sci Rep. 2015;5:16681.
13. Mosmann T. Rapid colorimetric assay for cellular growth and survival: application to proliferation and cytotoxicity assays. J Immunol Methods. 1983;65(1-2):55–63.
14. Labarca C, Paigen K. A simple, rapid and sensitive DNA assay procedure. Anal Biochem. 1980;102:344–52.
15. Gay RB, Bowers GN Jr. Optimum reaction conditions for human lactate dehydrogenase isoenzymes as they affect total lactate dehydrogenase activity. Clin Chem. 1968;14:740–53.
16. Rutkovskiy A, Stensløkken KO, Vaage IJ. Osteoblast differentiation at a glance. Med Sci Monit Basic Res. 2016;26(22):95–106.
17. Bessey OA, Lowry OH, Brock MJ. A method for the rapid determination of alkaline phosphatase with five cubic millimetres of serum. J Biol Chem. 1946;164:321–9.
18. Nakamura A, Dohi Y, Akahane M, Ohgushi H, Nakajima H, Funaoka H, Takakuracellular Y. Osteocalcin secretion as an early marker of in vitro osteogenic differentiation of rat mesenchymal stem cells. Tissue Eng Part C Methods. 2009;15(2):169–80.
19. Rosenberg N, Rosenberg O, Weizman A, Leschiner S, Sakoury Y, Fares F, Soudry M, Weisinger G, Veenman L, Gavish M. In vitro mitochondrial effects of PK 11195, a synthetic translocator protein 18 kDa (TSPO) ligand, in human osteoblast-like cells. J Bioenerg Biomembr. 2011;43(6):739–46.
20. Marom R, Shur I, Solomon R, Benayahu D. Characterization of adhesion and differentiation markers of osteogenic marrow stromal cells. J Cell Physiol. 2005;202(1):41–8.
21. Granéli C, Thorfve A, Ruetschi U, Brisby H, Thomsen P, Lindahl A, Karlsson C. Novel markers of osteogenic and adipogenic differentiation of human bone marrow stromal cells identified using a quantitative proteomics approach. Stem Cell Res. 2014;12(1):153–65.
22. Nagakura R, Yamamoto M, Jeong J, Hinata N, Yi K, Chang WJ, Abe S. Switching of Sox9 expression during musculoskeletal system development. Sci Rep. 2020;10:8425.
23. Ye J, Xiao J, Wang J, Ma Y, Zhang Y, Zhang Q. The interaction between intracellular energy metabolism and signaling pathways during osteogenesis. Front Mol Biosci. 2022;8:807487. www.frontiersin.org
24. Rosenberg N, Bettman N, Rosenberg O, Soudry M, Gavish M, Bar-Shalom R. Measurement of [18F]-fluorodeoxyglucose incorporation into human osteoblast—an experimental method. Cytotechnology. 2007;54(1):1–4.

25. Pauwels EKJ. Positron-emission tomography with [^{18}F]-fluorodeoxyglucose. Part I. Biochemical uptake mechanism and its implication for clinical studies. J Cancer Res Clin Oncol. 2000;126:549–59.
26. Golshani-Hebroni SG, Bessman SP. Hexokinase binding to mitochondria: a basis for proliferative energy metabolism. J Bioenerg Biomembr. 1997;29:331–8.
27. Brock CS, Meikle SR, Price P. Does fluorine–18 fluorodeoxyglucose metabolic imaging of tumours benefit oncology. Eur J Nucl Med. 1997;24:691–70.
28. Azarashvili T, Grachev D, Krestinina O, Evtodienko Y, Yurkov I, Papadopoulos V, Reiser G. The peripheral-type benzodiazepine receptor is involved in control of Ca2+-induced permeability transition pore opening in rat brain mitochondria. Cell Calcium. 2007;42(1):27–39.
29. Rosenberg N, Rosenberg O, Weizman A, Veenman L, Gavish M. In vitro catabolic effect of protoporphyrin IX in human osteoblast-like cells: possible role of the 18 kDa mitochondrial translocator protein. J Bioenerg Biomembr. 2013;45:333–41.
30. Wang L, Hsu H-Y, Xian C. Finite element analyses of vibration responses of an osteoblast in vitro. J Orthop Transl. 2016;7:82–5.
31. Nigg BM. Acceleration. In: Nigg BM, Herzog W, editors. Biomechanics of the musculo-skeletal system. 2nd ed. Chichester: Wiley; 1998. p. 300–1.
32. Oron U. Photoengineering of tissue repair in skeletal and cardiac muscles. Photomed Laser Surg. 2006;24:110–20.
33. Lipovsky A, Oron U, Gedanken A, Lubart R. Low-level visible light irradiation promotes proliferation of mesenchymal stem cells. Lasers Med Science. 2013;28(4):1113–7.
34. Higuchi A, Shen PY, Zhao JK, Chen CW, Ling QD, Chen H, Wang HC, Bing JT, Hsu ST. Osteoblast differentiation of amniotic fluid-derived stem cells irradiated with visible light. Tissue Eng Part A. 2011;17(21-22):2593–602.
35. Lewis JB, Wataha JC, Messer RLW, Caughman GB, Yamamoto T, Hsu SD. Blue light differentially alters cellular redox properties. J Biomed Mater Res B Appl Biomater. 2005;72(2):223–9.
36. Karu TI, Pyatibrat LV, Afanasyeva NI. A novel mitochondrial signaling pathway activated by visible to near infrared radiation. Photochem Photobiol. 2004;80(2):366–72.
37. Boyden ES, Zhang F, Bamberg E, Tsunoda SP, Mattis J, Yizhar O, Hegemann P, Deisseroth K. Millisecond-timescale, genetically targeted optical control of neural activity. Nat Neurosci. 2005;8(9):1263–8.
38. Rosenberg N, Gendelman R, Noofi N. Human osteoblast-like cells photobiomodulation *in vitro* by low intensity pulsed LED light. FEBS Open Bio. 2020;10(7):1276–87. https://doi.org/10.1002/2211-5463.12877.
39. Batabyal S, Cervenka G, Birch D, Kim Y, Mohanty S. Broadband activation by white opsin lowers intensity threshold for cellular stimulation. Sci Rep. 2015;5(17857):1–9.
40. McLeod KJ, Collazo L. Suppression of a differentiation response in MC-3T3-E1 osteoblast-like cells by sustained, low-level, 30 Hz magnetic-field exposure. Radiat Res. 2000;153:706–14.
41. Lohmann CH, Schwartz Z, Liu Y, Guerkov H, Dean DD, Simon B, Boyan BD. Pulsed electromagnetic field stimulation of MG63 osteoblast-like cells affects differentiation and local factor production. J Orthop Res. 2000;18(4):637–46.
42. Zhang X, Zhang J, Qu X, Wen J. Effects of different extremely low frequency electromagnetic fields on osteoblasts. Electromagn Biol Med. 2007;26(3):167–77.
43. Tong J, Sun L, Zhu B, Fan Y, Ma X, Yu L, Zhang J. Pulsed electromagnetic fields promote the proliferation and differentiation of osteoblasts by reinforcing intracellular calcium transients. Bioelectromagnetics. 2017;38(7):541–9.
44. Singh S, Kapoo N. Health implications of electromagnetic fields, mechanisms of action, and research needs. Adv Biol. 2014;2014:1–24.
45. Barnaba S, Papalia R, Ruzzini L, Sgambato A, Maffulli N, Denaro V. Effect of pulsed electromagnetic fields on human osteoblast cultures. Physiother Res Int. 2013;18(2):109–14.
46. Bullard EC. The secular change in the Earth's magnetic field. Geophys Suppl Month Not Royal Astronom Soc. 1948;5(7):248–57.
47. Rosenberg N, Rosenberg O, Soudry M. Pulsed low-intensity electromagnetic field (PEMF) affects cell cycle of human osteoblast-like cells *in vitro*. Am J Biomed Eng. 2012;2:181–4.

48. Rosenberg N. Triggering cultured human osteoblast-like cells' maturation by an extremely low magnitude alternating electromagnetic field. Iberoam J Med. 2020;3:12–7. https://doi.org/10.5281/zenodo.4319702.
49. Banes AJ, Link GW Jr, Gilbert JW, Tran Son Tay R, Monbureau O. Culturing cells in a mechanically active environment. Am Biotechnol Lab. 1990 May;8(7):12–22.
50. Pitsillides AA, Rawlinson SCF. Using cell and organ culture models to analyse responses of bone cells to mechanical stimulation. In: Helfrich MH, Ralston SH, editors. Bone research protocols. NJ: Humana Press; 2003. p. 399–422.
51. Colombo A, Cahill PA, Lally C. An analysis of the strain field in biaxial Flexcell membranes for different waveforms and frequencies. Proc Inst Mech Eng H. 2008;222(8):1235–45.
52. Rosenberg N, Levy M, Kyberd P, Kenwright J, Simpson H, Stein H. Experimental model for regulation of human osteoblast-like cells proliferation and metabolism *in vitro* by vibration. J Bone Joint Surg. 1998;1B(Supp 3):328.
53. Boal D. Chapter 10: Mechanics of the cell. In: Mechanical designs. Cambridge University Press; 2002. p. 335.
54. Nix Z, Kota D, Ratnayake I, Wang C, Smith S, Wood S. Spectral characterization of cell surface motion for mechanistic investigations of cellular mechanobiology. Prog Biophys Mol Biol. 2022;176:3–15. https://doi.org/10.1016/j.pbiomolbio.2022.08.002.
55. Rosenberg N, Soudry M. Chapter 26. Experimental *in vitro* methods for research of mechanotransduction in human osteoblasts. In: Poitout GD, editor. Biomechanics and biomaterials in orthopedics. 2nd ed. London: Springer; 2016. p. 311–7.
56. Rosenberg N, Rosenberg O, Halevi Politch J, Abramovich H. Optimal parameters for the enhancement of human osteoblast-like cell proliferation in vitro via shear stress induced by high-frequency mechanical vibration. Iberoam J Med. 2021;3(3):204–11.
57. Rosenberg N, Levy M, Francis M. Experimental model for stimulation of cultured human osteoblast-like cells by high frequency vibration. Cytotechnology. 2002;39:125–30.
58. Rosenberg N, Soudry M, Rosenberg O. Methods of cell-based bone regeneration—mini review. Recent Pat Biomed Eng. 2011;4:38–40. https://www.ingentaconnect.com/content/ben/biomeng/2011/00000004/00000001/art00006
59. Berkovich Y, Shapira J, Rosenberg N. Chapter 5. Mechanotransduction in osteoblasts. In: Rosenberg N, editor. Experimental methods for mesenchymal cell activation by biomechanical stimulation. Bentham Science; 2016. p. 49–57.
60. Rosenberg N. The role of the cytoskeleton in mechanotransduction in human osteoblast-like cells. Hum Exp Toxicol. 2003;22:271–4.
61. Rosenberg N, Rosenberg O. Extracorporeal human bone-like tissue generation. Bone Joint Res. 2012;1:1–7.
62. Cadossi R, Massari L, Racine Avila J, Aaron RK. Pulsed electromagnetic field stimulation of bone healing and joint preservation: cellular mechanisms of skeletal response. J Am Acad Orthop Surg Glob Res Rev. 2020;4:e19.00155.
63. Malone AMD, Anderson CT, Tummala P, Kwon RY, Johnston TR, Stearns T, Jacobs CR. Primary cilia mediate mechanosensing in bone cells by a calcium-independent mechanism. PNAS. 2007;104(33):13325–133330.

Determination of In Vitro Generated Bone Tissue

3

Current knowledge supports the conclusion that several factors can be combined to generate viable bone tissue in vitro. Osteoblasts can be manipulated in monolayer culture to induce proliferation, maturation, and matrix elaboration by applying alternating external biophysical stimulation with specific optimal parameters (frequency, amplitude) for each cellular activity. For this purpose, the bioreactor should contain cells seeded on an osteoconductive matrix and exposed to the optimal biophysical protocols.

The rationale for this approach is supported by the experimental evidence to be effective in mechanical stimulation setup [1] (Fig. 3.1).

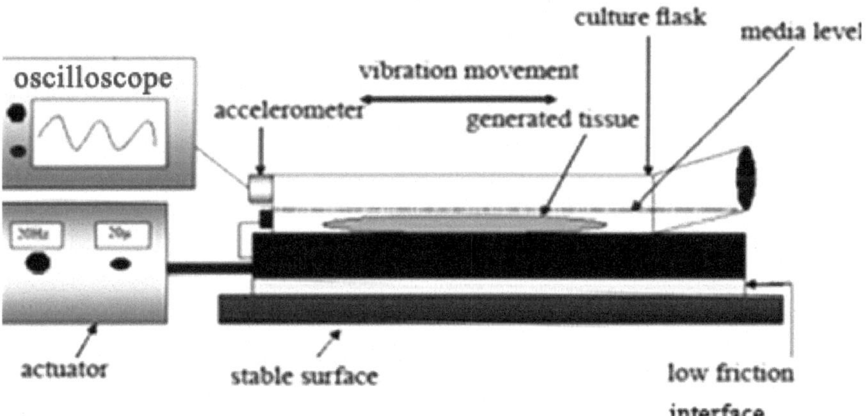

Fig. 3.1 Schematic presentation of mechanical stimulation of the generated tissue adherent to a plastic surface of culture flask by horizontal vibration. Sine-shaped vibration at 20 Hz frequency, 25–30 μm of displacement amplitude, and peak-to-peak acceleration of 0.5 m/s^2 (±0.1 m/s^2) was applied [1] (figure source—PhD Thesis authored by Rosenberg Nahum, "High-frequency alternating biophysical stimulation of human osteoblast," School of Pharmacy and Biomedical Sciences, Faculty of Science, University of Portsmouth, October 2021)

N. Rosenberg, *Biophysical Osteoblast Stimulation for Bone Grafting and Regeneration*, https://doi.org/10.1007/978-3-031-06920-8_3

By using this experimental approach, following the culture confluency, the cells are passaged and seeded onto three-dimensional granules of tricalcium phosphate (βTCP, diameter 0.5 mm, pores 300–500 μ, 5 ml total volume), cultured in osteogenic media (similar to the described above for culturing of primary explant cultures of osteoblasts) and exposed to the mechanical conditions (30 Hz of vibration, 25 μm of displacement) for 2 min every 24 h for 21 days. At the end of this period, bone-like tissue (BLT) is generated (Fig. 3.2) via endochondral-like osteogenesis pathway (Fig. 3.3) [1].

The porous βTCP, calcium sulfate, or hydroxyapatite can be used as an osteoconductive milieu for the in vitro three-dimensional bone matrix generation by the

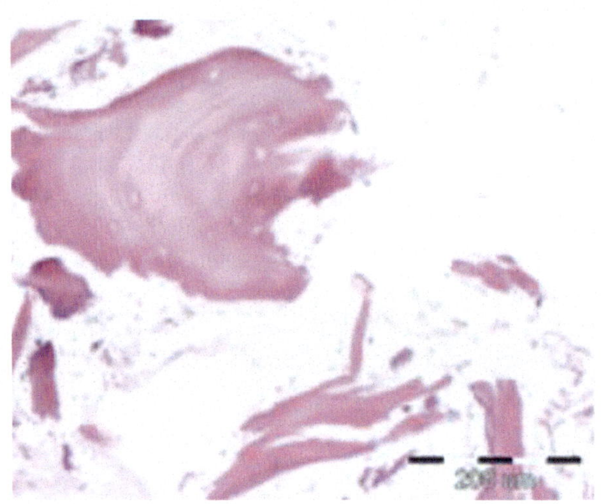

Fig. 3.2 Sample micrograph of the generated tissue after 2 weeks of culture showing organized bone-like areas (Hematoxylin and Eosin stain) [1] (figure source—PhD Thesis authored by Rosenberg Nahum, "High-frequency alternating biophysical stimulation of human osteoblast," School of Pharmacy and Biomedical Sciences, Faculty of Science, University of Portsmouth, October 2021)

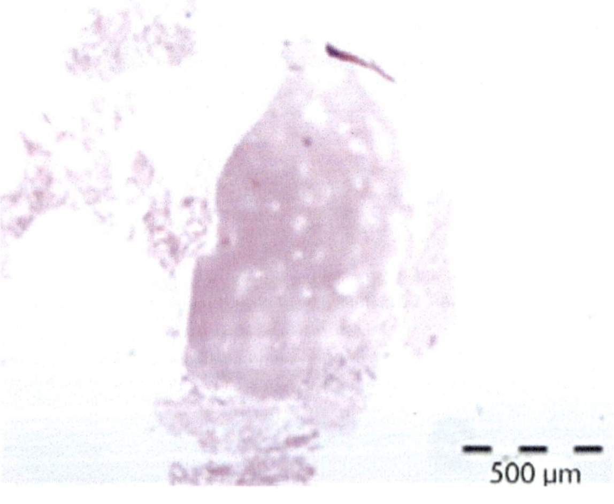

Fig. 3.3 Micrograph showing cartilaginous tissue generating bone-like tissue presented in Fig. 3.2 following initial 7 days of incubation (Hematoxylin and Eosin stain) [1] (figure source—PhD Thesis authored by Rosenberg Nahum, "High-frequency alternating biophysical stimulation of human osteoblast," School of Pharmacy and Biomedical Sciences, Faculty of Science, University of Portsmouth, October 2021)

osteoblast due to their proven experimental and clinical osteointegration effect [2] that is already proven to be efficient for an osseointegration coverage of metal endoprostheses [3, 4] (Fig. 3.4).

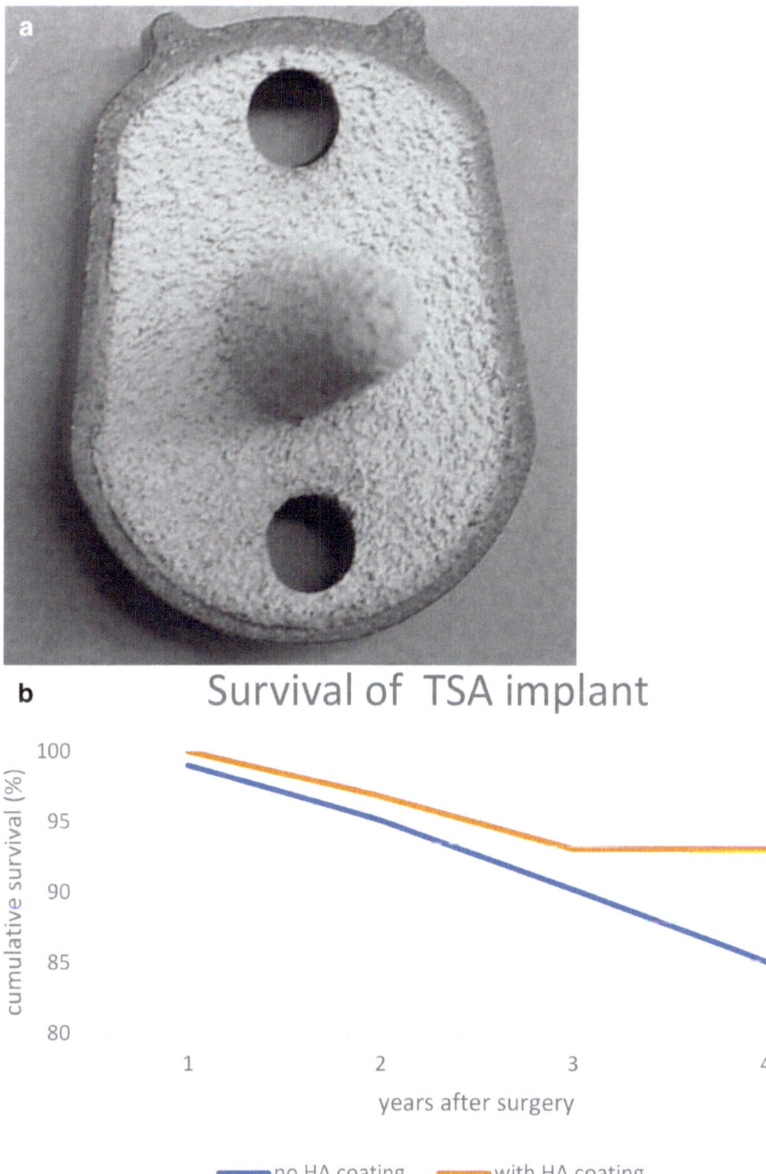

Fig. 3.4 (**a**) Baseplate of the glenoid component of shoulder replacement endoprosthesis. This baseplate is covered by porous hydroxyapatite on its interface with bone and showed improved osseointegration. (**b**) Short-term survivorship of Total Shoulder Arthroplasty (TSA) is significantly higher when the only change of the prosthetic design is the cover of the glenoid component by the hydroxyapatite (HA) in the implant–bone interface (shown in **a**) [3]

Osteoblast penetration into a porous inorganic scaffold requires an interconnected structure of pores with a minimum diameter of 200 μm [5] (Fig. 3.5).

Numerous attempts exist to design additional synthetic scaffold materials with enhanced osseointegration due to the ability to control their stiffness properties, such as polyethylene glycol, polydimethylsiloxane, vinyl polymers, polyesters, polyesters, polyacrylamide, and polyurethane [6]. An additional scaffold type that can be considered for effective osteoblast integration for bone generation is the nickel-titanium shape memory metal alloy Nitinol (NiTi) (Figs. 3.6 and 3.7) [7].

Modulating surface stiffness properties for the adherent cells, out of the known optimal range of 25–40 MPa for osteoblastogenesis, might reveal improvement in the generation of bone-like tissue [8].

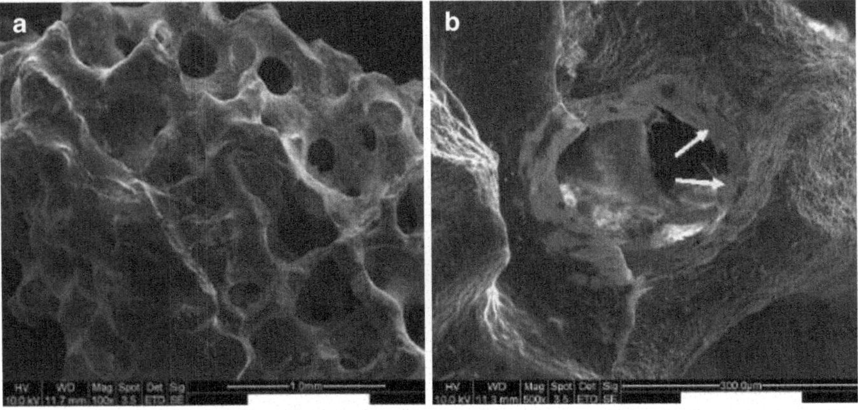

Fig. 3.5 (**a**) Scanning electron microscopy (SEM) micrograph of a porous βTCP construct. (**b**) SEM micrograph of osteoblasts (white arrows) attached to a porous βTCP construct

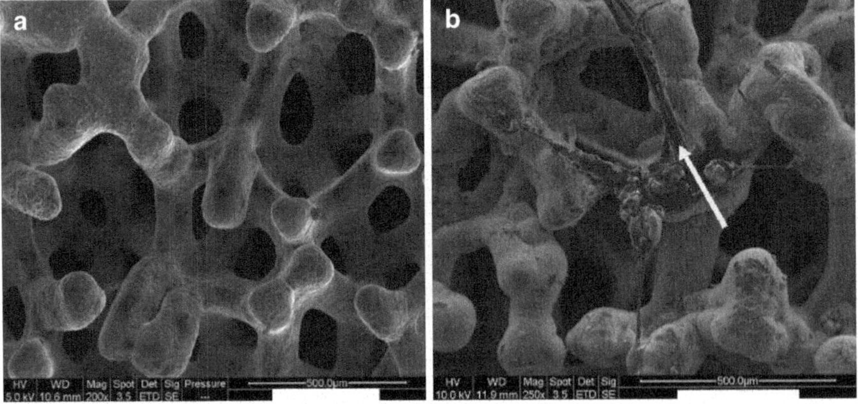

Fig. 3.6 (**a**) SEM micrograph of porous NiTi construct. (**b**) SEM micrograph of porous NiTi construct with adherent osteoblast that elaborates matrix (arrow)

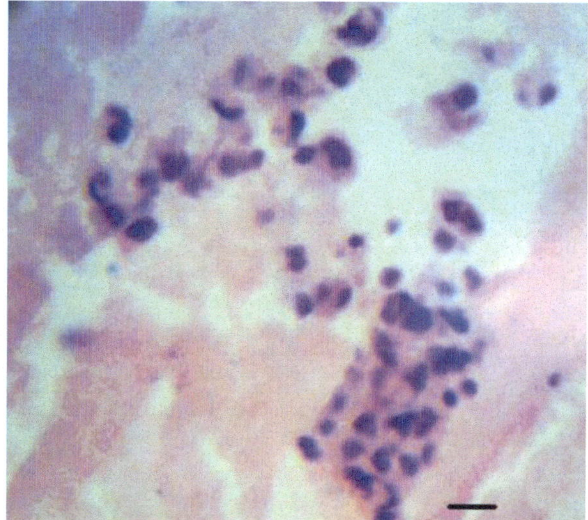

Fig. 3.7 Exfoliative cytology image of osteoblasts scraped from the surface of retrieved titanium alloy bone implant. HE staining, scale—50 μm

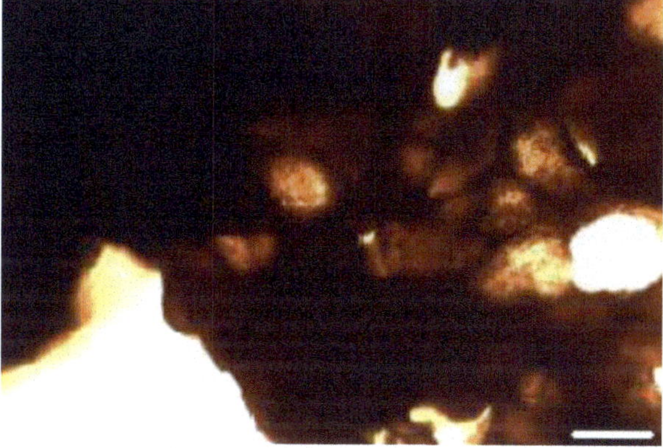

Fig. 3.8 Microscopic image of the β-TCP porous construct after seeding with osteoblasts and cultured in osteogenic media according to the mechanical protocol with exposure to a controlled vibration in the 30–70 Hz range. The interconnected pores are filled with an inorganic matrix generated by the osteoblasts. Scale 200 μm

The experiments with a porous βTCP scaffold have shown a promising ability to generate viable bone tissue in vitro [1, 9]. Following these experiments, the in vitro generated bone-like tissue (BLT) characteristics, safety, and efficiency to close critical bone gaps in vivo were investigated in mice and rat models. The alternating mechanical stimulation with precise parameters is essential for generating vital bone tissue in vitro for implantation as a bone graft. This protocol of seeding bone marrow-derived human primary osteoblast-like cells (>70% positive for osteocalcin [10]) on clinically approved β-tricalcium phosphate (β-TCP) matrix with exposure to effective mechanical stimulation inside a bioreactor was found to be effective to engineer in vitro bone graft-like material. The generated by the osteoblasts inorganic matrix fills the interconnecting pores in the β-TCP osteoconductive matrix (Fig. 3.8) The bioengineered end-product mimics live autologous bone graft in terms of three-dimensional organization, bone critical gaps closure capability, and functional integration with host tissue [9] (Fig. 3.9).

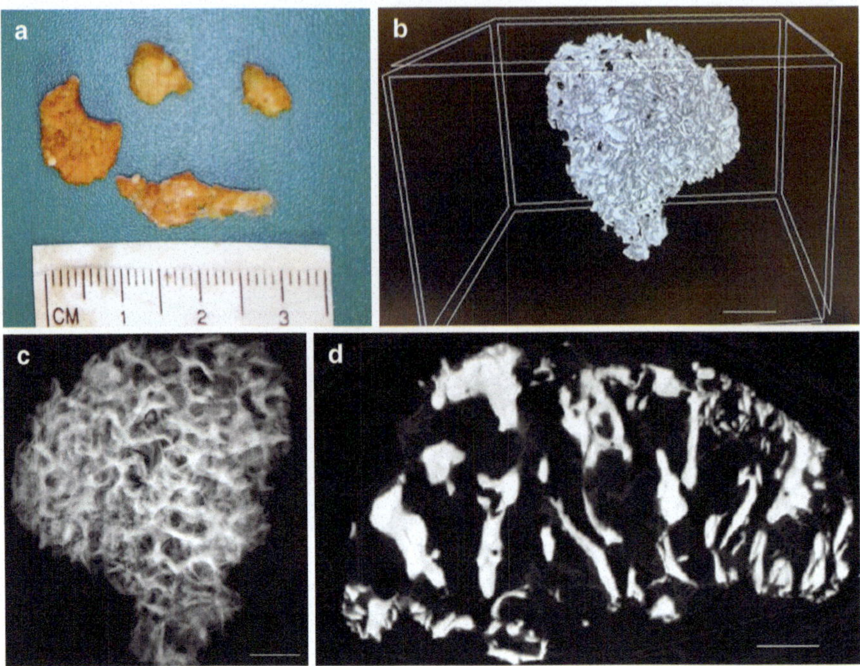

Fig. 3.9 (**a**) Gross appearance of the in vitro generated bone-like material. (**b**, **c**) Three-dimensional reconstructed micro-CT images of the in vitro generated bone. (**d**) Transverse micro-CT images of the in vitro generated bone. The trabecular structure of the generated bone is evident. Scale—2 mm

The characterization of the generated material as a bone-like tissue is supported by its positive immunohistochemical staining for the bone markers (osteocalcin and collagen I) (Fig. 3.10).

The generated by the described above method BLT is potentially safe for clinical use, following the evidence of the lack of cytotoxicity (tested by a concomitant culture of mouse fibroblast cells L929 using MTT cell viability assay) and no evidence of tumorigenicity potential (tested in a NOD-SCID mice model and showed no signs of morbidity, mortality or histopathological evidence of malignancy in the general and local examination, 2 months following BLT subcutaneous inoculation) [9].

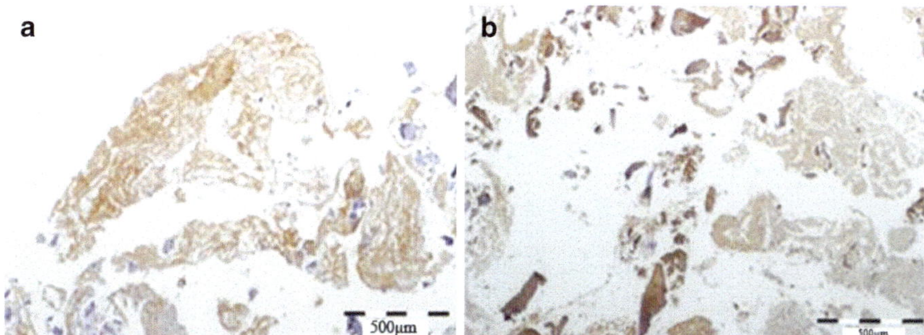

Fig. 3.10 Micrographs showing immunohistochemical staining for collagen 1 (**a**) and osteocalcin (**b**) in the experimentally generated BLT [9] (figure source—PhD Thesis authored by Rosenberg Nahum, "High-frequency alternating biophysical stimulation of human osteoblast," School of Pharmacy and Biomedical Sciences, Faculty of Science, University of Portsmouth, October 2021)

References

1. Rosenberg N, Rosenberg O. Extracorporeal human bone-like tissue generation. Bone Joint Res. 2012;1:1–7.
2. Luvizuto ER, Queiroz TP, Margonar R, Panzarini SR, Hochuli-Vieira E, Okamoto T, OkamotoR. Osteoconductive properties of β-tricalcium phosphate matrix, polylactic and polyglycolic acid gel, and calcium phosphate cement in bone defects. J Craniofac Surg. 2012;23(5):e430–3.
3. Rosenberg N, Neumann L, Modi A, Mersich IJ, Wallace AW. Improvements in survival of the uncemented Nottingham Total Shoulder prosthesis: a prospective comparative study. BMC Musculoskelet Disord. 2007;8(76):1–11.
4. Jalota S, Bhaduria SB, Tas AC. Osteoblast proliferation on neat and apatite-like calcium phosphate-coated titanium foam scaffolds. Mater Sci Eng C. 2007;27(3):432–40.
5. Lim TC, Chian KS, Leong KF. Cryogenic prototyping of chitosan scaffolds with controlled micro and macro architecture and their effect on in vivo neo-vascularization and cellular infiltration. J Biomed Mater Res. 2010;94a(4):1303–11. https://doi.org/10.1002/jbm.a.32747.
6. El-Rashidy AA, El Moshy S, Radwan IA, Rady D, Abbass MMS, Dörfer CE, Fawzy KM, El-Sayed. Effect of polymeric matrix stiffness on osteogenic differentiation of mesenchymal stem/progenitor cells: concise review. Polymers. 2021;13:2950.
7. Kapanenab A, Ilvesaroab J, Danilova A, Ryhänenc J, Lehenkaric P, Tuukkanen J. Behaviour of Nitinol in osteoblast-like ROS-17 cell cultures. Biomaterials. 2002;23(3):645–50.
8. Witkowska-Zimny M, Walenko K, Wrobel E, Mrowka P, Mikulska A, Przybylski J, et al. Cell Biol Int. 2013;37(6):608–16.
9. Rosenberg N, Rosenberg O. Safety and efficacy of *in vitro* generated bone-like material for *in vivo* bone regeneration—a feasibility study. Heliyon. 2020;6(1):1–7.e03191. https://doi.org/10.1016/j.heliyon.2020.e03191.
10. Rosenberg N, Soudry M, Rosenberg O, Blumenfeld I, Blumenfeld Z. The role of activin A in the human osteoblast cell cycle: a preliminary experimental *in vitro* study. Exp Clin Endocrinol Diabetes. 2010;118:708–12.

The Osseointegration Potential of Engineered Bone-Like Tissue μm

4

Thus, bone-like viable tissue can be generated in vitro by combining an inorganic matrix, osteoblasts, osteogenic media, and applying adequate biophysical (mechanical) stimulation of the osteoblasts. Furthermore, this tissue has no toxic or tumorigenic potential. The primary use of such tissue is to bridge critical gaps in bone, i.e., to be used as efficient bone graft material.

4.1 The Bone-Like Tissue for In Vivo Implantation

To test the osseointegration capacity of the Bone-Like Tissue (BLT) in vivo and the ability of the implanted tissue to interact with the host bone at the orthotopic site, the in vitro *generated* BLT was implanted into a rat calvaria critical-size defect model [1]. The ability of the BLT to promote critical-size defect healing was monitored during the experiment using X-ray imaging (Fig. 4.1) and at the experimental endpoint using histological evaluation by HE staining (Fig. 4.2).

The histological evaluation included a gap closure with woven bone score as a marker for newly in situ developed bone tissue based on the following gradation: Grade 0 = 0% closure; Grade 1 = up to 25% closure; Grade 2 = up to 50% closure; Grade 3 = up to 75% closure; and Grade 4 = up to 100% closure. Since nonhealing bone wounds generally result in a thin collagenous fibrous tissue within the defect with no visible bone ingrowth [2], the amount of fibrous tissue was scored based on the following gradation: (−) no fibrosis; (+) marginal; (++) minor; and (+++) pronounced. If any, the length of the remaining gap was measured under the microscope using a calibration scale.

Subsequentially, the results showed that the critical gaps in BLT-implanted animals (experimental model with rats) demonstrated full bridging of the calvaria critical bone gap with vascularized woven bone (Fig. 4.2).

N. Rosenberg, *Biophysical Osteoblast Stimulation for Bone Grafting and Regeneration*, https://doi.org/10.1007/978-3-031-06920-8_4

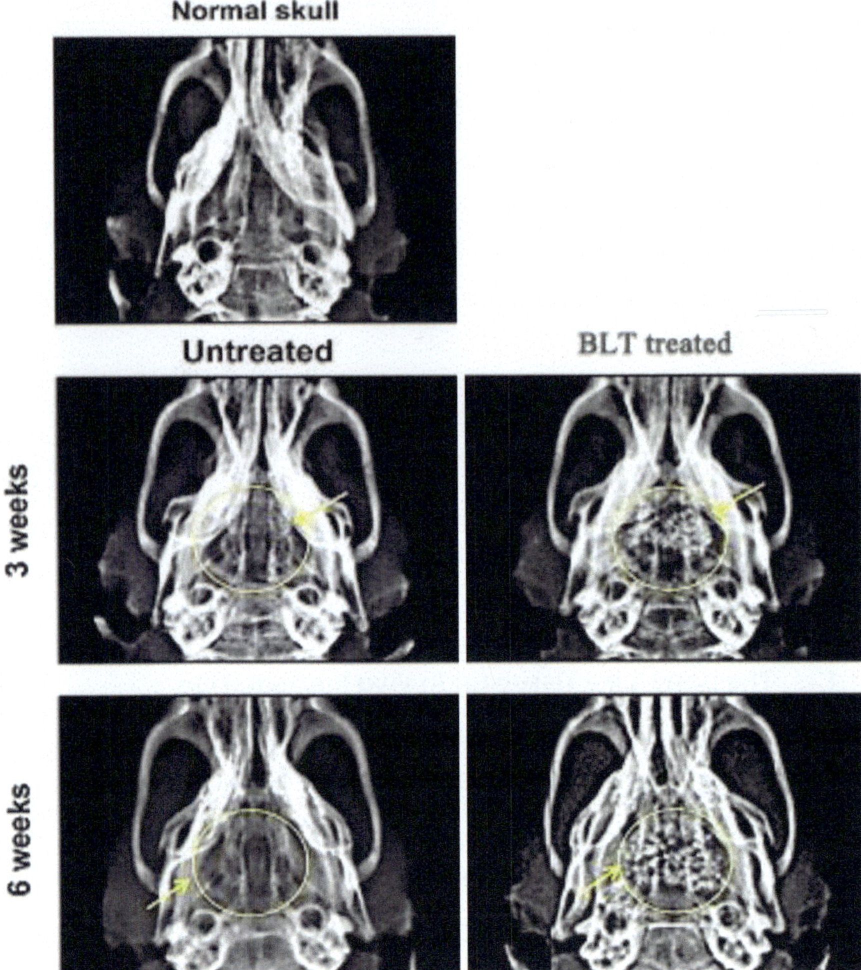

Fig. 4.1 X-ray radiographs of BLT-treated and untreated rats bearing critical-size calvarial defects. The rats were imaged at the 3rd and 6th postoperative weeks. The circle and the arrows indicate the original circular-shaped gap in the calvarial bone. A normal skull radiograph was taken from the unoperated and untreated rats for comparison

Therefore, with the accomplishment of this series of research projects and according to the initial hypothesis, the specific alternating high-frequency (in the range of 20–70 Hz) biophysical parameters were found to be effective in inducing proliferation and phenotypic cell function of human osteoblasts in vitro. It became apparent that by implementing the biophysical cell stimulation using mechanical vibration, the viable bone tissue could be generated in vitro, which is safe and effective for potential use as a bone graft in vivo.

There is additional promising evidence that similarly to the described above in the rat animal model, the critical bone gap can be bridged by a nanofiber polylactic

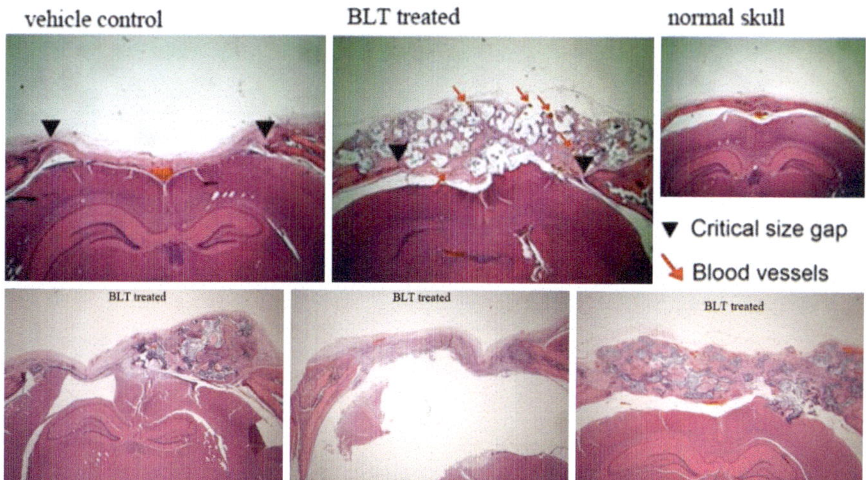

Fig. 4.2 Representative histological photomicrographs of calvarial defects at the 6th postoperative week. HE-stained preparations of controls and BLT-treated rats upon completing 6 weeks of the experiment. Black triangles indicate the borders of the original gap prior to implantation. Long arrows point to newly formed blood vessels inside the BLT [1] (figure source—PhD Thesis authored by Rosenberg Nahum, "High-frequency alternating biophysical stimulation of human osteoblast," School of Pharmacy and Biomedical Sciences, Faculty of Science, University of Portsmouth, October 2021)

glycolic acid scaffold seeded by MSCs originating from peripheral blood [3]. This fact provides an even more versatile source for the cells that can propagate to osteoblastic lineage for the generation of BLT in vitro for implantation as a bone graft in vivo.

References

1. Rosenberg N, Rosenberg O. Safety and efficacy of *in vitro* generated bone-like material for *in vivo* bone regeneration—a feasibility study. Heliyon. 2020;6(1):1–7.e03191. https://doi.org/10.1016/j.heliyon.2020.e03191.
2. Cowan CM, Shi YY, Aalami OO, Chou YF, Mari C, Thomas R, Quarto N, Contag CH, Wu B, Longaker MT. Adipose-derived adult stromal cells heal critical-size mouse calvarial defects. Nat Biotechnol. 2004;22:560–7.
3. Wu G, Pan M, Wang X, Wen J, Cao S, Li Z, Li Y, Qian C, Liu Z, Wu W, Zhu L, Guo J. Osteogenesis of peripheral blood mesenchymal stem cells in self assembling peptide nanofiber for healing critical size calvarial bony defect. Sci Rep. 2015;5:16681. https://doi.org/10.1038/srep16681.

5 The Clinical Potential of the In Vitro Generated Bone-Like Tissue

Bone has a high potential to regenerate after damage; however, the efficacious repair of large defects resulting from resection, trauma, extended fractures, or inadequately vascularized bone gaps will not achieve satisfactory healing and will be considered non-union [1] (Fig. 5.1).

The definition of non-union is a failure of the fracture to heal in 6 months in a patient in whom progressive repair had not been observed radiographically between the third and sixth month after the fracture. Therefore, up to 10% of patients will experience a non-union fracture and require the implantation of bone grafts. In addition, bone grafting is also required at the sites of bone deficiency because of surgical interventions such as failed endoprosthesis surgery and craniotomy. Cumulatively, *ca.* 500,000 surgical cases of bone grafting procedures occur annually in the USA [2]. Notably, the demand for bone grafts is expected to be even greater over the next decade as the population ages and life expectancy increases.

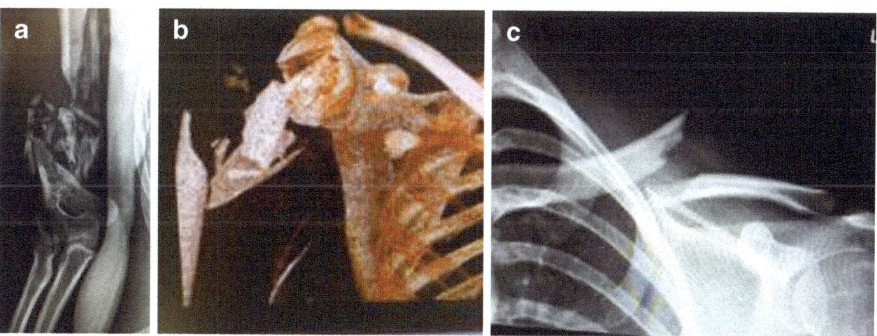

Fig. 5.1 Examples (on radiographs imaging) of bone fractures that are prone to the development of non-union (presented by radiographs. (**a**) Comminuted and displaced fracture of the proximal humerus. (**b**) Comminuted and displaced fracture of the distal humerus. (**c**) Displaced fracture of the clavicle

N. Rosenberg, *Biophysical Osteoblast Stimulation for Bone Grafting and Regeneration*, https://doi.org/10.1007/978-3-031-06920-8_5

In a clinical setting, most bone grafting relies on bone graft replacement materials (e.g., calcium phosphate matrices and collagen sponges) and natural bone grafts (e.g., autografts, allografts, and xenografts). However, disadvantages such as donor site morbidity (autografts), limited availability (autografts), risk of rejection (allografts/xenografts), risk of infection (allografts/xenografts), and ethical issues (xenografts) call for novel therapeutic options. Recent clinical trials in bone regeneration provide a comprehensive example of the trend toward exploring cell-based therapy as an alternative to natural bone grafts. Personalized and functional tissue-engineered bone grafts may provide an unlimited source to overcome chronic shortages in natural grafts.

The target cells for bone tissue engineering are differentiated osteoblasts derived from the pool of mesenchymal stem cells present in the bone marrow. Mesenchymal stem cells (MSCs) are pluripotent and can differentiate into various cell types such as osteoblasts, chondrocytes, muscle cells, adipocytes, or fibroblasts. To achieve high purity of osteoblasts culture, the osteogenic differentiation media is usually supplemented with transforming growth factor-beta, vitamin C, and dexamethasone [1, 3]. The differentiation process into the osteoblast lineage can be monitored by the appearance of a specific gene expression profile, including the presence of mRNA for RUNX2, *alkaline phosphatase, osteopontin, osteonectin, osteocalcin*, or the proteins themselves. Cells displaying this phenotype can be further injected directly into the bone defect or first seeded onto a scaffold and transferred into the bone defect within the scaffold. Implantation of both a scaffold and cells can have an additive positive effect [4].

Thus, the main promising insight, as described in the previous chapter, is that the in vitro generated viable BLT, according to the biophysical principles, is safe for implantation in vivo and capable of bridging critical gaps in bone in vivo [5, 6]. This ability is reduced without the utilization of the biophysical stimulation of osteoblasts. Therefore, this became possible because of the mechanical stimulation setup. This unique finding presents exciting potential for the clinical use of the generated in vitro BLT as an autologous bone graft.

5.1 Translation from Basic Research to Clinical Practice

Bone grafts materials should demonstrate three major interrelated properties for optimal bone regeneration [7]:

(i) *Osteoconduction*—supporting the attachment of new osteoblasts and osteoprogenitor cells.
(ii) *Osteoinduction*—induction of osteoprogenitor cells (or other non-differentiated cells) to differentiate toward osteoblast lineage.
(iii) *Osteogenesis*—formation of new bone via osteoblasts-derived osteoid that subsequently mineralizes.

Although synthetic bone grafts, either with or without exogenous growth factors, provide osteoconductive and osteoinductive microenvironment, only autologous grafts meet all three criteria for bone regeneration and are considered the gold standard for bone implantation [1]. Therefore, disadvantages such as donor site morbidity (autografts), limited availability (autografts), risk of rejection (allografts/xenografts), risk of infection (allografts/xenografts), and ethical issues (xenografts) call for novel therapeutic options. Implantation of bone concentrated marrow aspirate, which contains osteogenic progenitors, in the attempt to treat sizeable critical bone gaps, has not yet shown sufficient clinical success because only a small fraction of implanted MSCs in the aspirate survive after implantation [8]. Similarly, the potential osteogenic solid effect of the recombinant bone morphogenic proteins (BMPs), planted directly or on an osteoconductive inorganic carrier, is not widely introduced for clinical use as the treatment of fracture non-union or for bridging the critical bone gaps because of the potentially devastating effect of neoplastic transformation [9].

In general, personalized and functional tissue-engineered bone grafts may provide an unlimited source to overcome the chronic shortage in natural grafts while addressing all three mentioned optimal criteria.

Accordingly, future experimental projects should be planned to further understand the biophysical manipulation of human osteoblasts.

5.2 The Perspective for the Clinical Use of in Vitro-Generated Bone-Like Material

The ongoing research projects aiming to evaluate the osteoblast response to biophysical stimulation should culminate in clinically efficient methods for generating an unlimited source of autologous bone grafting material. Thus, these bioengineering techniques will be used in the framework of a personalized approach for bone grafting with the most clinically efficient bone graft. The logical plan is to proceed with this research according to the following topics:

- Improvement of the ability to grow osteoblast monolayer cultures originating from the peripheral blood samples with low content of MSCs [10]. This will simplify the process of the eventual in vitro bone generation, according to the principles described above, without the need for bone marrow sampling, which is a more invasive procedure.
- Construction of a versatile apparatus that can provide a combined tuneable application of low friction mechanical vibration, electromagnetic fields, and LED light irradiance to the osteoblast cultures. This device will allow an application of a combined biophysical effect on the cultured osteoblasts.
- The important limitation of the ability to record the precise kinetics of the generated BLT in culture, when it is only partially adherent to the plastic surface, from where a piezoelectric sensor records the movement, is the source of uncertainty on the exposure to the exact spectrum of harmonic frequencies of movement.

The suspension in the culture media BLT is affected by fluid and container movements. Technically, it is very difficult to measure the force frequencies on the BLT by external sensors in the presented experimental setup. Therefore optical recording of live cells and the BLT in culture in static conditions and during external mechanical stimulation for growth, proliferation, and kinetics of external biophysical manipulation should give more precise data without interfering with culture conditions. For this purpose, the Digital inline-holographic microscope (DIHM) technique might be utilized [11]. The DIHM method is based on a 5-megapixel complementary metal oxide semiconductor (CMOS) image sensor and a multicolor LED illumination capable of video imaging recording of cell cultures and media with different spectral absorbances. This allows high flexibility in applying direct measurements of the suspended in the culture media BLT kinetics.

- Comparison of the gene expression in the cultured osteoblasts following exposure to the biophysical modalities described above, e.g., the DNA microarray methods, should give a substantial indication of the uniform pathways for cellular response to the alternating mechanical forces and electromagnetic fields.
- Modulation of surface stiffness properties for the adherent cells, out of the known optimal range of 25–40 MPa for osteoblastogenesis, might reveal improvement in the generation of the bone-like tissue generated.
- The accomplishment of the preclinical studies of safety and efficacy on a large animal model (similar to the presented above studies on murine models) of the use of the in vitro generated bone-like material by the implementation of biophysical stimulation of osteoblasts in vitro, and subsequentially preparation of the described method for the clinical studies, following application for the approval from the regulatory authorities. This will be the desired translational step from the basic research of biophysical (mechanical and electromagnetic) stimulation of cells to induce bone matrix elaboration in vitro toward the clinical use as an autologous bone graft.

5.3 Conclusion

Human osteoblasts react to external biophysical stimulation by several mechanisms, including propagating an external signal by electrical currents via a cellular membrane and intracellular biochemical secondary pathways. The resultant effect is the modulation of proliferation and phenotypic activity. There is evidence from a series of research projects that there are specific alternating high-frequency biophysical parameters for the induction of the phenotypic cell function of human osteoblasts in vitro and that by using the specific biophysical parameters for osteoblast stimulation, viable bone tissue can be generated in vitro that is safe and effective for use as a bone graft in vivo. This evidence is supported by experiments utilizing specially designed modalities with the external application of mechanical vibration, alternating LED irradiance, and electromagnetic fields in the 20–70 Hz range of frequencies to monolayer cultures of human osteoblasts. By using these methods, effective biophysical parameters for cell stimulation were defined. These parameters are different and distinctive for phenotypic cell function vs. synthetic activities. Furthermore, the experimental evidence indicates that osteoblast photobiomodulation occurs at a low-intensity threshold of 40 Hz 0.04 W/m^2 pulsed irradiance, and cell exposure to pulsed electromagnetic field (PEMF) at a distinct range of 5–15 kHz of basic frequency in pulses of 20–30 Hz caused a shift of the cell cycle toward the G1 Phase. External mechanical stimulation of cells by vibration showed different sub-ranges of effective vibration parameters for osteoblast proliferation and phenotypic cell function, i.e., 60 and 20 Hz frequencies.

Thus, these experiments indicate that applying alternating biophysical energy (light, electromagnetic field, mechanical vibration) in the frequency range of 20–70 Hz causes proliferative and phenotypic effects in human osteoblasts.

Subsequentially, by applying biophysical stimulation, viable bone-like tissue can be generated in vitro that is safe and effective for bridging critical bone gaps in vivo (investigated using small animal models).

Following the series of research projects, the high-frequency ranges of alternating biophysical stimulation for phenotypic cell function and proliferation of human osteoblasts in vitro were determined. These findings reveal the ability to implement in vitro tissue engineering techniques of osteoblast manipulation in culture by external biophysical methods for clinical use to treat critical bone loss by autologous bone grafting. Accordingly, biophysical stimulation of cells is a way to generate bone-like tissue in vitro for the potential clinical implementation as autologous bone graft in orthopedic, neuro, maxillofacial, and dental surgery.

References

1. Alexander PG, Hofer HR, Clark KL, Tuan RS. Mesenchymal stem cells in musculoskeletal tissue engineering. In: Lanza R, Langer R, Vacanti J, editors. Principles of tissue engineering. 4th ed. Waltham, MA: Elsevier; 2014. p. 1171–200.
2. Greenwald AS, Boden SD, Goldberg VM, Khan Y, Laurencin CT, Rosier RN. Bone-graft substitutes: facts, fictions, and applications. J Bone Joint Surg Am. 2001;83(Suppl 2 Pt 2):98–103.
3. Park JB. The effects of dexamethasone, ascorbic acid, and beta-glycerophosphate on osteoblastic differentiation by regulating estrogen receptor and osteopontin expression. J Surg Res. 2012;173:99–104.
4. Olender E, Brubaker S, Uhrynowska-Tyszkiewicz I, Wojtowicz A, Kaminski A. Autologous osteoblast transplantation, an innovative method of bone defect treatment: role of a tissue and cell bank in the process. Transplant Proc. 2014;46:2867–72.
5. Rosenberg N, Rosenberg O. Safety and efficacy of *in vitro* generated bone-like material for *in vivo* bone regeneration—a feasibility study. Heliyon. 2020;6(1):1–7.e03191. https://doi.org/10.1016/j.heliyon.2020.e03191.
6. Cowan CM, Shi YY, Aalami OO, Chou YF, Mari C, Thomas R, Quarto N, Contag CH, Wu B, Longaker MT. Adipose-derived adult stromal cells heal critical-size mouse calvarial defects. Nat Biotechnol. 2004;22:560–7.
7. Mercado-Pagan AE, Stahl AM, Shanjani Y, Yang Y. Vascularization in bone tissue engineering constructs. Ann Biomed Eng. 2015;43:718–29.

8. Hernigou P, Poignard A, Manicom O, Mathieu G, Rouard H. The use of percutaneous autologous bone marrow transplantation in nonunion and avascular necrosis of bone. J Bone Joint Surg Br. 2005;87(7):896–902.
9. Sheikh Z, Javaid MA, Hamdan N, Hashmi R. Bone Regeneration Using Bone Morphogenetic Proteins and Various Biomaterial Carriers. Materials (Basel). 2015;8(4):1778–816.
10. Liangliang X, Li G. Circulating mesenchymal stem cells and their clinical implications. J Orthop Translat. 2014;2:1–7.
11. Scholz G, Mariana S, Dharmawan AB, Syamsu I, Hörmann P, Reuse C, Hartmann J, Hiller K, Prades JD, Wasisto HS, Waag A. Continuous live-cell culture imaging and single-cell tracking by computational lensfree LED microscopy. Sensors (Basel). 2019;19(5):1234.

Part II

Clinical Applications of Biophysical Stimulation for Bone Regeneration

6 Local Vibration for Fracture Healing

As described previously, human osteoblast proliferation and metabolic activity in vitro are enhanced by mechanical stimulation by high-frequency low-intensity vibration [1]. The logical clinical application of such a biological phenomenon is to enhance bone fracture healing by applying this stimulus to the bone fracture site. This approach was investigated in animal experiments and showed promising results [2]. Unfortunately, this theoretical method has not reached a clinical implementation yet.

Effective transfer of the external mechanical vibration with the optimal mechanical parameters is essential to provide optimal mechanical stimulation to the osteoblasts in the fracture site. Therefore, this method might be best for fractures in bones in direct proximity to the skin. For this purpose, wrist and ankle fractures are potential candidates for this type of treatment.

The ability to enhance fracture healing and bridge fracture nonunion without surgical intervention might partially resolve enormous morbidity in patients with bone fractures and a significant burden in health services. Even in highly susceptible to spontaneous union, fractures of the humerus may reach 30% of the nonunion or delayed union rate [3].

Distal radius and ulna fractures are common; up to 2.5% of hospital emergency visits are attributed to this pathology [4]. Currently, in physically active young patients with displaced fractures, a surgical approach by open reduction and internal is indicated. Still, in older patients with comminuted fractures due to osteoporosis or in young patients with un-displaced fractures, conservative treatment by external plaster immobilization is usually preferred. In the latter group of conservatively treated patients, additional physiotherapy treatment is required following the immobilization cast removal to restore the good wrist and hand functions. If a shorter time of immobilization, until fracture healing, could be achieved, faster and better rehabilitation results are anticipated. For this purpose, the development and construction of a method and device for stimulation of bone fracture union by external mechanical stimulation based on vibration parameters for stimulation of osteoblasts for bone matrix elaboration, to bridge the traumatic gap in bone, is feasible, according to the experimental evidence described above, and of very high clinical importance.

N. Rosenberg, *Biophysical Osteoblast Stimulation for Bone Grafting and Regeneration*, https://doi.org/10.1007/978-3-031-06920-8_6

A special device for this purpose should apply vibrations on the healing bone. Such a device might be composed of an electro-vibrating actuator or a piezoelectric actuator. The actuator will be placed in a hole created in the immobilizing POP cast. According to animal studies, it can be controlled for the effective biomechanical protocol, especially in the frequency range of 35–50 Hz [2]. The shorter time until roentgenological evidence of fracture union that is expected to be achieved faster than usual minimal 6 weeks, will be considered as clinical success of the method.

The main concern about the effectiveness of such a concept in clinical use is a possible distortion of the external mechanical force that propagates via soft tissue (skin, muscles, tendons, etc.). But from the ex vivo investigation (unpublished data) on the animal (porcine) wrist, by using a mechanical actuator attached to the skin (Fig. 6.1), it appears that the external signal is accurately transferred into the medullary canal of radius without change in amplitude and frequency of the vibration.

The previous evidence supports this observation that alternating mechanical signal, by ultrasonic pressure wave, applied to a cortical bone gap effectively propagates through the trabecular bone, which is embedded in a bone marrow fluid milieu [5] (Figs. 6.2 and 6.3).

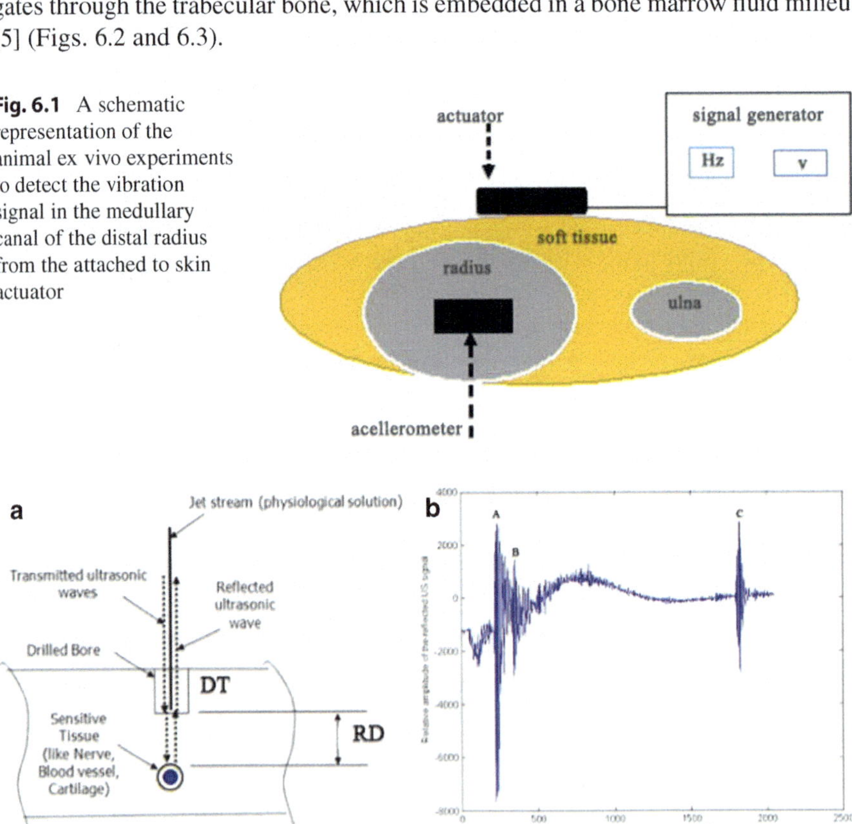

Fig. 6.1 A schematic representation of the animal ex vivo experiments to detect the vibration signal in the medullary canal of the distal radius from the attached to skin actuator

Fig. 6.2 Schematic representation of the ultrasound wave propagation in trabecular bone via water jet and bone marrow and its detection as a reflected wave. (**a**) *DT* drilled tract in cortical bone. *RD* residual distance consisting of trabecular bone, up to opposite cortex or dense trabecular bone. (**b**) An example of a reflected sonographic pattern: A–B distance represents the DT, B–C distance represents the RD. The C point represents the reflection from the opposite dense bone

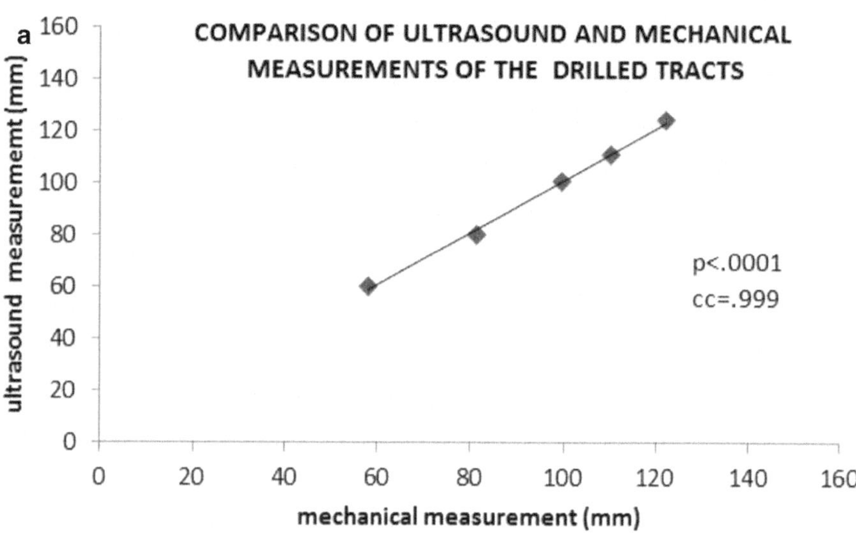

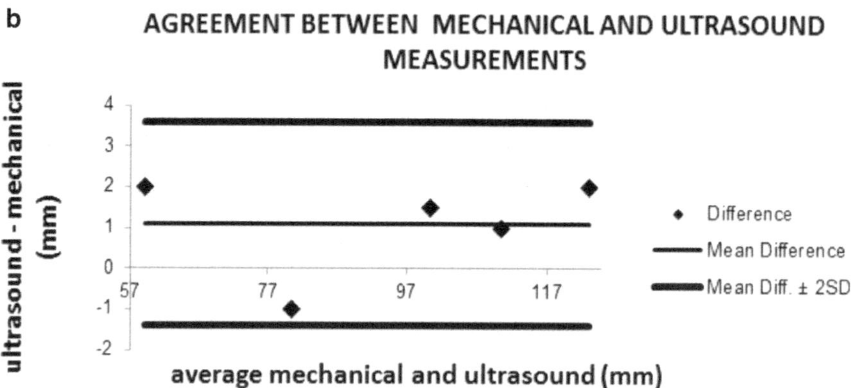

Fig. 6.3 (**a**) High correlation between mechanical and ultrasound measurements in bone. Correlation coefficient −0.999, $p < 0.0001$. (**b**) Mean difference between these methods of measurements is 1.1 mm, SD 1.24 mm (by Bland Altman plot)

Thus, the experimental data on the ability of external biophysical stimulation of osteoblast supports a promising potential to evolve into clinical use for several common types of bone fractures using relatively simple mechanical devices.

References

1. Rosenberg N, Levy M, Francis M. Experimental model for stimulation of cultured human osteoblast-like cells by high frequency vibration. Cytotechnology. 2002;39:125–30.
2. Wang J, Leung KS, Chow SKH, Cheung WH. The effect of whole body vibration on fracture healing—a systematic review. Eur. Cells Mater. 2017;34:108–27.

3. Naclerio EH, McKee MD. Approach to humeral shaft nonunion: evaluation and surgical techniques. JAAOS. 2022;30(2):50–9.
4. Larsen CF, Lauritsen J. Epidemiology of acute wrist trauma. Int J Epidemiol. 1993;22(5):911–6.
5. Rosenberg N, Halevy-Politch J. Intraosseous monitoring of drilling in lumbar vertebrae by ultrasound: an experimental feasibility study. PLoS One. 2017;12(5):e0174545.

Whole-Body Vibration Enhances Bone Regeneration: For Treatment of Osteoporosis

7

Under physiologic conditions, human skeletal muscles contract spontaneously at the frequency of around 20 Hz, which subsequently induces ongoing mechanical stimulation of adjacent bones [1, 2]. This vibrational/mechanical loading is clinically utilized to enhance bone strength and mass [3] and was found to facilitate bone mass production, fracture healing, and osseointegration of bone-anchored implants [4, 5]. There is ample preclinical in vivo evidence supporting the utilization of mechanical stimulation for improved bone homeostasis, healing, and osseointegration of implants [6–9]. At the in vitro level, it has been repeatedly demonstrated that short-term vibrational stimulation in the range of 20–40 Hz facilitates differentiation toward and maturation of osteoblast as evaluated using increased alkaline phosphatase activity, matrix synthesis, and mineralization [10, 11, 7, 12, 13, 14]. Altogether, mechanical stimuli position themselves as a valuable, effective, and straightforward strategy to promote osteogenic differentiation [15].

Thus, according to the experimental evidence from the in vitro studies described in the previous section, bone mass can be potentially increased by stimulating the osteoblast matrix elaboration following alternating mechanical stimulation in the high-frequency, low-magnitude/intensity range (LIV). To remind: high frequency is in the range of 20–60 Hz, and low magnitude is when the acceleration is below 9.8 m/s^2 (g). This might be an effective nonpharmacological method for treating osteoporosis [16]. Clinton T. Rubin initially popularized the therapeutic potential of LIV for osteoporosis [17]. Still, space flight agencies have created even prior interest in this therapeutical modality since the early 1960s, aiming to prevent the catabolic effects on bone tissue due to the lack of gravity environment.

The device for exposing the patient to whole-body vibration usually consists of a vibrating platform with support to hands while standing. This setup causes propagation of the vibration movements to the body via hands and feet, which are attached to the vibrating surface (Fig. 7.1).

Currently, the research results show that different protocols of LIV enhance the bone mineral density, especially in women in postmenopausal age when the risk for

N. Rosenberg, *Biophysical Osteoblast Stimulation for Bone Grafting and Regeneration*, https://doi.org/10.1007/978-3-031-06920-8_7

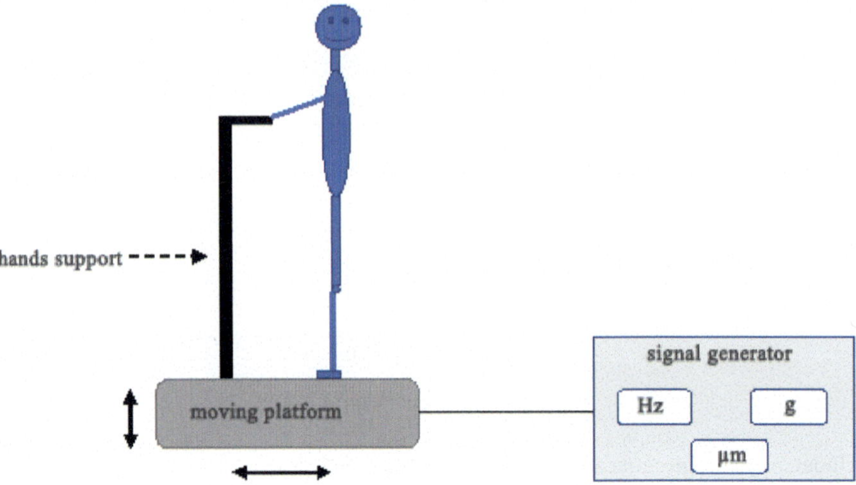

Fig. 7.1 Schematic representation of the generic concept of whole-body vibration for increasing bone mass. The vibration movement is transferred from the supporting platform via feet and hands

osteoporosis is high and causes significant morbidity. This effect is also found in children and young adults, suggesting the possible enhanced MSCs maturation, because the mesenchymal progenitors are more abundant in younger ages [18, 19]. The side effects, including damage to joints and muscles, are not frequent if the vibration's peak-to-peak acceleration does not exceed 1 g [16]. The lack of significant information from controlled research and inconsistent protocols of timing and precise parameters that are most efficient in increasing bone mass are the primary causes of the poor rate of adaptation of LIV for the treatment of osteoporosis. To mention that studies in vitro (discussed in Part I) showed that even short (2 min) exposure of cultured osteoblasts to vibration in the range of 20–60 Hz affects their synthetic and proliferative activity [10]. This fact might indicate a possible optimal mechanical protocol for LIV. Therefore, there is a promising future for this type of biophysical stimulation of osteoblast in a clinical setting, mainly because it is expected to reduce the need for pharmacological intervention in treating osteoporosis. Future well-designed clinical studies should resolve this uncertainty.

References

1. Nigg BM. Acceleration. In: Nigg BM, Herzog W, editors. Biomechanics of the musculoskeletal system. 2nd ed. Chichester: Wiley; 1998. p. 300–1.
2. Behrens SB, Deren ME, Monchik KO. A review of bone growth stimulation for fracture treatment. Curr Orthop Pract. 2013;24:84–9.
3. Larsen CF, Lauritsen J. Epidemiology of acute wrist trauma. Int J Epidemiol. 1993;22(5):911–6.
4. Torcasio A, van Lenthe GH, Van Oosterwyck H. The importance of loading frequency, rate and vibration for enhancing bone adaptation and implant osseointegration. Eur Cell Mater. 2008;16:56–68.

5. Ulstrup AK. Biomechanical concepts of fracture healing in weight-bearing long bones. Acta Orthop Belg. 2008;74:291–302.
6. Jing D, Tong S, Cai J, Zhai M, Shen G, Wang X, Luo E, Luo Z. Mechanical vibration mitigates the decrease of bone quantity and bone quality of leptin receptor-deficient db/db mice by promoting bone formation and inhibiting bone resorption. J Bone Miner Res. 2016;31(9):1713–24.
7. Zhou Y, Guan X, Liu T, Wang X, Yu M, Yang G, Wang H. Whole body vibration improves osseointegration by up-regulating osteoblastic activity but down-regulating osteoblast-mediated osteoclastogenesis via ERK1/2 pathway. Bone. 2015;71:17–24.
8. Qing F, Xie P, Liem YS, Chen Y, Chen X, Zhu X, Fan Y, Yang X, Zhang X. Administration duration influences the effects of low-magnitude, high-frequency vibration on ovariectomized rat bone. J Orthop Res. 2015;34(7):1147–57.
9. Jing D, Tong S, Zhai M, Li X, Cai J, Wu Y, Shen G, Zhang X, Xu Q, Guo Z, Luo E. Effect of low-level mechanical vibration on osteogenesis and osseointegration of porous titanium implants in the repair of long bone defects. Sci Rep. 2015;5:17134.
10. Rosenberg N, Levy M, Francis M. Experimental model for stimulation of cultured human osteoblast-like cells by high frequency vibration. Cytotechnology. 2002;39:125–30.
11. Rosenberg N, Rosenberg O. Extracorporeal human bone-like tissue generation. Bone Joint Res. 2012;1:1–7.
12. Kim IS, Song YM, Lee B, Hwang SJ. Human mesenchymal stromal cells are mechanosensitive to vibration stimuli. J Dent Res. 2012;91:1135–40.
13. Pre D, Ceccarelli G, Gastaldi G, Asti A, Saino E, Visai L, Benazzo F, Cusella De Angelis MG, Magenes G. The differentiation of human adipose-derived stem cells (hASCs) into osteoblasts is promoted by low amplitude, high frequency vibration treatment. Bone. 2011;49:295–303.
14. Wu XT, Sun LW, Qi HY, Shi H, Fan YB. The bio-response of osteocytes and its regulation on osteoblasts under vibration. Cell Biol Int. 2016;40:397–406.
15. Gelinsky M, Lode A, Bernhardt A, Rösen-Wolff A. Stem cell engineering for regeneration of bone tissue. In: Artmann GM, Minger S, Hescheler J, editors. Stem cell engineering: principles and applications. Berlin: Springer; 2011. p. 383–99.
16. Thompson WR, Yen SS, Rubin j. Vibration therapy: clinical applications in bone. Curr Opin Endocrinol Diabetes Obes. 2014;21(6):447–53.
17. Rubin CT, McLeod KJ. Promotion of bony ingrowth by frequency-specific, low-amplitude mechanical strain. Clin Orthop Relat Res. 1994;298:165–74.
18. Weber-Rajek M, Mieszkowski J, Niespodziński B, Ciechanowska K. Whole-body vibration exercise in postmenopausal osteoporosis. Prz Menopauzalny. 2015;14(1):41–7.
19. Dionello CF, Sá-Caputo D, Pereira HVFS, Sousa-Gonçalves CR, Maiworm AI, Morel DS, Moreira-Marconi E, Paineiras-Domingos LL, Bemben D, Bernardo-Filho M. Effects of whole body vibration exercises on bone mineral density of women with postmenopausal osteoporosis without medications: novel findings and literature review. J Musculoskelet Neuronal Interact. 2016;16(3):193–203.

Distraction Osteogenesis

8

Bone is generated and remodeled along the vector of external mechanical force from gravity and muscle contraction. This phenomenon, known as "Wolff's law" [1], reflects the interconnected responses of the BMU components (osteoblasts, osteoclasts, endothelial cells) to static tension; when osteoblast elaborate bone matrix, osteoclasts resorb it according to tension vector and endothelial cells induce angiogenesis, which is essential for the overall bone maintenance as a live organ. The experimental model for the research of constant tension application on osteoblasts is usually composed of cells exposed to continuous fluid flow in a closed loop (Fig. 8.1) [2]. Thus, the external shear stress of the media flow causes strain on cultured cells adherent to a solid surface [3]. This method allows a uniform magnitude of shear stress on cells to be controlled and tuned by generating a laminar flow of media over the cells in a monolayer. Indeed, experiments following this principle show the osteogenic effect in cultured osteoblasts, e.g., induction of Nitric Oxide release, which enhances the proliferation, maturation, and survival of osteoblasts with parallel inhibition of osteoclasts, therefore inducing overall bone formation [4, 5]. This cellular response occurs by different pathways following constant or alternating

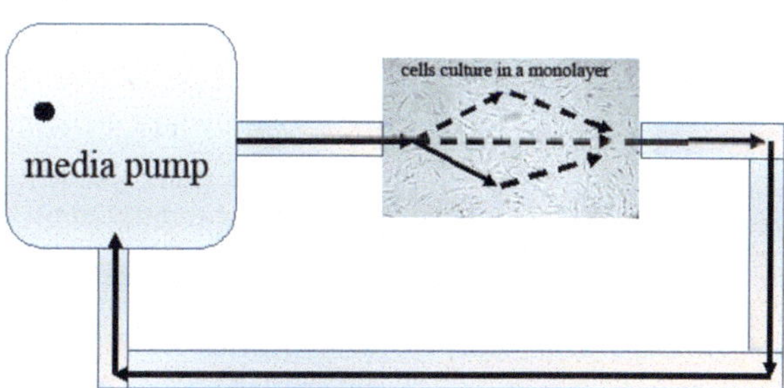

Fig. 8.1 Schematic representation of the experimental method for application of constant shearing force on cells adherent to solid surface by unidirectional laminar flow of fluid

N. Rosenberg, *Biophysical Osteoblast Stimulation for Bone Grafting and Regeneration*, https://doi.org/10.1007/978-3-031-06920-8_8

external shearing stress; therefore, there is a reason to hypothesize that the osteogenesis by distraction is mediated by different cellular machinery from discussed in the previous chapters, by high-frequency low magnitude alternating biophysical stimulation, i.e., G-protein calcium-dependent pathway for alternating stimulation and G-protein calcium-independent pathway for the constant mechanical stress.

But much before the experimental support, the intuitive impression that bone can be elongated by continuous external distraction force over osteotomized area was translated into efforts for clinical use. The purpose of such a procedure is to correct bone deformities. This concept was initially proposed in 1869 by German surgeon Bernhard von Langenbeck and gradually evolved from traction via the extremities edges, then unilateral external fixators, by circular frames, and distraction over intramedullary rods, which are regulated externally by electromagnetic motors [6].

The principle of the distraction osteogenesis by either method is similar—generation of reduced resistance of the bone cortex (osteotomy or corticotomy) with preservation of local blood supply and periosteum, which are the sources of MSCs, and application of distraction force on the weakened bone site where some stability exists due to preserved trabecular bone. The initial osteotomy causes local hematoma that recruits the cellular and humoral components required for further bone regeneration (the latency phase), then 5–7 days later, the distraction is commenced with the pace of 1–2 mm/day that generates the buildup of connective tissue, according to the direction of the applied external tension, which becomes a scaffold for osteoblast activity for osteogenesis up to the formation of bone trabeculae, composed of woven bone (distraction phase). According to the preoperative plan, the process involves distraction, usually not more than 20% of the treated bone length, to avoid serious damage to soft tissue and adjacent joints [7]. The regenerative tissue undergoes remodeling by osteoclasts with the eventual mature bone formation with adequate stability for weight-bearing (consolidation phase). This phase lasts at least 8–12 weeks with the subsequential removal of the external distraction/stabilization device [8].

The initial attempts of traction of the limb edge and traction by external fixator on both sides of the corticotomy site by one-dimensional external fixator were inefficient because of the lack of three-dimensional stability and control of the distraction site in the bone.

The breakthrough came by using circular frame stabilization around the corticotomy site, allowing the three-dimensional stabilization and control of the distraction pace and the direction of the distraction force (Fig. 8.2). Ilizarov GA invented and popularized this method. Since the middle 1950s, it gradually became a gold standard distraction osteogenesis technique with extensive implementation worldwide [9].

The Ilizarov method based on the use of a circular fixation frame eventually evolved into a more controllable Taylor Spatial frame that operates hinges between

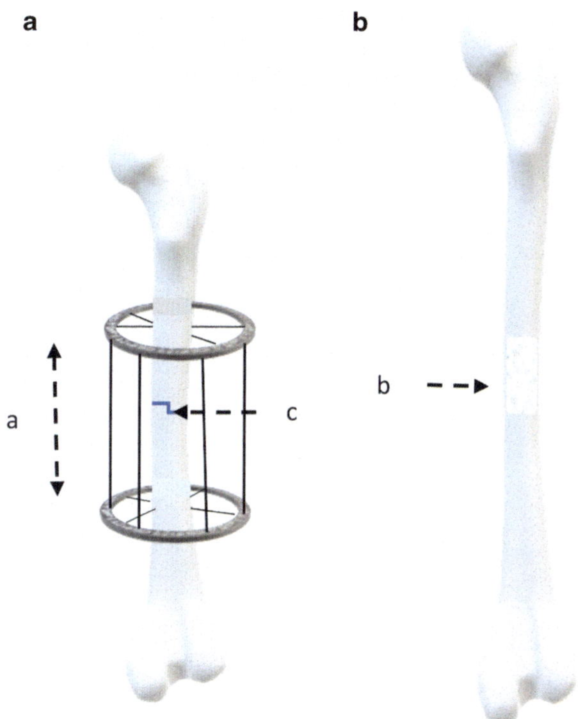

Fig. 8.2 Simplistic description of distraction osteogenesis clinical utilization. (**a**) Corticotomy or osteotomy is generated (c) after the treated bone is stabilized by a circular external fixation frame. According to the clinical protocol, the osteotomized site is distracted via the external frame. Eventually, new bone is generated in the distraction site (b), and the bone is elongated into the desired shape (**b**)

the fixation rings with the ability of three-dimensional control. These hinges are designed according to the principle of Stewart platform that allows six degrees of freedom. Furthermore, this method allows the computerized perioperative planning for distraction protocol for the improved three-dimensional control of bone elongation and angular correction, according to the same distraction osteogenesis principles [10].

Due to the serious and unresolved complication of the distraction osteogenesis technique [11], either by the circular or the circular spatial frames, especially because of high risk for the damage to adjacent soft tissue and joints, the natural evolvement of the method via intramedullary expansible rode is desirable. Tor the optimal outcome of the distraction osteogenesis over the intramedullary nail, external control of the distraction mechanism is advantageous. The conceptually preferred method for this purpose is the magnetically controlled telescopic growing nails [12]. The latter method is still in the evolutionary process, and if it proves to cause improved functional outcomes in patients, it might become the chosen clinical method of distraction osteogenesis (Fig. 8.3).

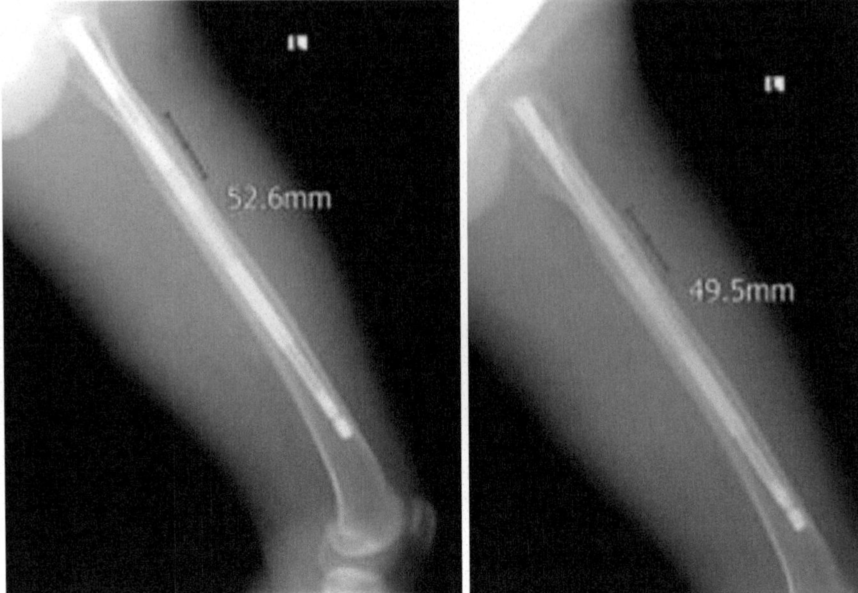

Fig. 8.3 Example radiographs of femur elongated by 5 cm along the externally controlled telescopic intramedullary road. The patient was treated for short stature

References

1. Wolff J. The law of bone remodeling. Berlin: Springer; 1986. (translation of the German 1892 edition).
2. Leiser Y, Rachmiel A. Cellular molecular pathways in distraction osteogenesis, chapter 7. In: Rosenberg N, editor. Mesenchymal cell activation by biomechanical stimulation and its clinical prospective. Sharjah UAE: Bentham Sciemce; 2015. p. 65–71.
3. Reich KM, Gay CV, Frangos JA. Fluid shear stress as a mediator of osteoblast cyclic adenosine monophosphate production. J Cell Physio. 1990;143(1):100–4.
4. McAllister TN, Frangos JA. Steady and transient fluid shear stress stimulate NO release in osteoblasts through distinct biochemical pathways. J Bone Miner Res. 1999 Jun;14(6):930–6.
5. Nascimento MHM, Pelegrino MT, Pieretti JC, Seabra AB. How can nitric oxide help osteogenesis? AIMS Mol Sci. 2020;7(1):29–48.
6. Hosn GA. Limb lengthening history, evolution, complications and current concepts. J Orthop Traumatol. 2020;21:3.
7. Domingos G, Armés H, Dias I, Viegas C, Requicha J. Distraction osteogenesis: biological principles and its application in companion animals. In: Barbeck M, Rosenberg N, Rider P, Kačarević ŽP, Jung O, editors. Clinical implementation of bone regeneration and maintenance. London: Intechopen; 2019. p. 93–103.
8. Natu SS, Ali I, Alam S, Giri KY, Agarwal A, Kulkarn VA. The biology of distraction osteogenesis for correction of mandibular and craniomaxillofacial defects: a review. Dent Res J (Isfahan). 2014;11(1):16–26.
9. Spiegelberg B, Parratt T, Dheerendra SK, Khan WS, Jennings R, Marsh DR. Ilizarov principles of deformity correction. Ann R Coll Surg Engl. 2010;92(2):101–5.

10. Keshet D, Eidelman M. Clinical utility of the Taylor spatial frame for limb deformities. Orthop Res Rev. 2017;9:51–61.
11. Papakostidis C, Bhandari M, Giannoudis PV. Distraction osteogenesis in the treatment of long bone defects of the lower limbs: effectiveness, complications and clinical results; a systematic review and meta-analysis. Bone Joint J. 2013;95-B(12):1673–80.
12. Lee DH, Kim S, Lee JW, Park H, Kim TY, Kim HW. A comparison of the device-related complications of intramedullary lengthening nails using a new classification system. Biomed Res Int. 2017;2017:8032510.

Evolving Clinical Modalities for Bone Regeneration by Biophysical Stimulation

9

9.1 Electromagnetic Stimulation for Bone Fracture Healing

The application of electromagnetic field (EMF) in a clinical setup has a therapeutic value in enhancing damaged bone regeneration by finetuning osteoblast proliferation and maturation. Therefore, EMF is used to treat delayed union or non-union in bone fractures. This method evolved from in vitro (see Part I) and animal studies to an established clinical modality, which is currently approved by the regulatory authorities [1].

The initial successful attempt to treat fracture non-union by the constant electrical current of 10 μampere, applied directly into the non-union site in medial ankle malleolus, described by Friedenberg, Harlow, and Brighton [2], brought to the general public interest (Donald Janson. The New York Times 1971, October 30, p. 29) because it showed a potential for the noninvasive method for treatment of bone fractures.

According to the same rationale for the effectiveness of alternating mode of biophysical application in the 10–20 Hz range described above, the method of electromagnetic treatment of fracture non-union was directed for the use of pulsed EMF (PEMF) in this frequency range. PEMF with bursts frequency of 15 Hz and 1 kHz basic frequency of each burst, 50 mV amplitude, causes 76–79% success in non-union treatment when the non-union gap is below 10 mm [3]. Additional studies support this observation, but not in all [4]. Thus, there is a good indication of the clinical effectiveness of PEMF in the treatment of fracture non-union and delayed unions. However, there is still some uncertainty about the most effective protocols for PEMF use, and this issue probably prevents the routine use of this type of treatment. It is clear from the in vitro studies that the triggering values for osteoblast proliferation are in the range of 0.2–0.4 mT and 10 Hz of PEMF [5] but the preferable parameters for clinical use are still elusive and should be further verified in the future studies to support the widespread use of PEMF in fracture treatment.

N. Rosenberg, *Biophysical Osteoblast Stimulation for Bone Grafting and Regeneration*, https://doi.org/10.1007/978-3-031-06920-8_9

9.2 Fracture Healing by Ultrasound Stimulation

Ultrasound source causes acoustic mechanical energy propagation by pressure waves via fluid milieu and produces excitation of particles (at frequencies above 20 kHz). Thus, ultrasound wave energy causes shearing stress on cells and generates local heat according to its intensity, similar to other forms of mechanical energy application. Therefore, it is logical to assume that it might be used as a modality of biophysical stimulation of cells. These appealing characteristics of ultrasound ignited its use in stimulating bone regeneration via a biophysical effect on osteoblasts. The ultrasound waves are transferred effectively in the trabecular bone, embedded by fatty bone marrow, and reflected by the dense cortical bone (Fig. 6.2) [6]. Therefore, fluid communication is essential for the ultrasound wave to access the trabecular bone. Such a situation exists in bone fractures; therefore, using ultrasound stimulation for bone regeneration to treat bone fractures is logical and appealing to avoid surgical intervention for fracture fixation. The prerequisite for this type of use of ultrasound is avoidance of excessive heat generation that might damage the cells and extracellular matrix. Therefore, the ultrasound should be of low intensity, i.e., below 3 W/cm^2 [7].

According to these principles, the low-intensity pulsed ultrasound (LIPUS) method evolved to treat bone fractures, especially delayed union or non-union of fractures. The rationale for this treatment is the evidence that LIPUS induces the previously described mechanotransduction effects via cellular ion channels, i.e., differentiation of MSCs into the osteoblastic lineage and subsequential osteogenesis via Rho-associated kinase-Cot/Tpl2-MEK-ERK signalling pathway and RUNX2 and Osteocalcin mRNAs [8, 9].

Initial large-scale clinical reports of the use of 30 mW/cm^2 intensity LIPUS showed a 38% reduced fracture healing time and a range of 70–85% of non-union healing without thermal damaging side effects [10]. According to this evidence, the LIPUS was approved for clinical use by the regulatory authorities (FDA).

Since the exact optimal treatment protocols are still not sufficiently clear, the LIPUS has not reached significantly widespread clinical use yet. However, its promising advantage as a sole or adjuvant treatment modality in fracture treatment, acute or delayed, is not disputable.

9.3 Fracture Healing by Radiofrequency (RF) Electromagnetic Fields

An additional biophysical modality that has been suggested to have an enhancive effect on bone generation is pulsed radiofrequency (RF) electromagnetic fields. This suggestion is based on the research impression in the early studies that RF PEMF might be efficient for healing soft tissue lesions [11].

RF spectrum ranges between 3 kHz and 300 GHz [12]. The main concern about its clinical use is the increased thermic effect that can damage the viability of bone.

But exposure to RF EMF up to 570 W × s/cm^2 is not causing bone necrosis [13]. Furthermore, 250 mW 2100 MHz EMF application causes improved fracture healing in mandibular bone in an animal model via osteoblast activation [14].

Clinical evidence of the efficiency of RF EMF to induce bone regeneration is still scarce [15]. Although some clinical evidence exists, the substantiality of this method should be based on well-designed clinical studies that might be anticipated in the future.

References

1. Cadossi R, Massari L, Racine-Avila J, Aaron RK. Pulsed electromagnetic field stimulation of bone healing and joint preservation: cellular Mechanisms of skeletal response. J Am Acad Orthop Surg Glob Res Rev. 2020;4(5):e1900155.
2. Friedenberg ZB, Harlow MC, Brighton CT. Healing of nonunion of the medial malleolus by means of direct current: a case report. J Trauma. 1971;11(10):883–5.
3. Punt BJ, den Hoed PT, Fontijne WPJ. Pulsed electromagnetic fields in the treatment of non-union. Eur J Orthop Surg Traumatol. 2008;18:127–33.
4. Aleem IS, Aleem I, Evaniew N, Busse JW, Yaszemski M, Agarwal A, Einhorn T, Bhandari M. Efficacy of electrical stimulators for bone healing: a meta-analysis of randomized sham-controlled trials. Sci Rep. 2016;19(6):31724.
5. Barnaba S, Papalia R, Ruzzini L, Sgambato A, Maffulli N, Denaro V. Effect of pulsed electromagnetic fields on human osteoblast cultures. Physiother Res Int. 2013;18(2):109–14.
6. Rusnak I, Rosenberg N, Halevy-Politch J. Trabecular bone attenuation and velocity assess by ultrasound pulse-echoes. Appl Acoust. 2020;157:107007.
7. Nicholson JA, Tsang STJ, MacGillivray TJ, Perks F, Simpson AHRW. What is the role of ultrasound in fracture management? Bone Joint Res. 2019;8(7):304–12.
8. Tan Y, Guo Y, Reed-Maldonado AB, Li Z, Lin G, Xia S-J, Lue TF. Low-intensity pulsed ultrasound stimulates proliferation of stem/progenitor cells: what we need to know to translate basic science research into clinical applications. Asian J Androl. 2021;23:602–10.
9. Kusuyama J, Bandow K, Shamoto M, Kakimoto K, Ohnishi T, Matsuguchi T. Low intensity pulsed ultrasound (LIPUS) influences the multilineage differentiation of mesenchymal stem and progenitor cell lines through ROCK-Cot/Tpl2-MEK-ERK signaling pathway. J Biol Chem. 2014;289(15):10330–44.
10. Romano CL, Romano D, Logoluso N. Low-intensity pulsed ultrasound for the treatment of bone delayed union or nonunion: a review. Ultrasound Med Biol. 2009;35(4):529–36.
11. Lightwood R. Radio energy healing-medicine or myth? J Med Eng Technol. 1977;1(4):231.
12. D'Andrea JA, Ziriax JM, Adair ER. Radio frequency electromagnetic fields: mild hyperthermia and safety standards. Prog Brain Res. 2007;162:107–35.
13. Menendez M, Ishihara A, Weisbrode S, Bertone A. Radiofrequency energy on cortical bone and soft tissue: a pilot study. Clin Orthop Relat Res. 2010;468(4):1157–64.
14. Durgun M, Dasdag S, Erbatur S, Yegin K, Durgun SO, Uzun C, Ogucu G, Alabalik U, Akdag MZ. Effect of 2100 MHz mobile phone radiation on healing of mandibular fractures: an experimental study in rabbits. Biotechnol Biotechnol Equip. 2016;30(1):112–20.
15. Sharp IK. Stimulation of bone union by externally applied radio-frequency energy. Injury. 1983;14(6):523–30.

To Summarize 10

Different biophysical sources can induce osteoblast for matrix elaboration and bone generation by altering trans-membranous electric currents. The biophysical energy can be mechanical or electromagnetic, with similar but not identical activation of cellular pathways. But the parameters of biophysical stimulation vary from constant to alternating, and the alternating can be of different frequencies to elicit different cell responses, i.e., proliferation, synthetic activity, and cell death. This type of response of osteoblast to external energy can be and already is used for clinical application for the treatment of different pathological conditions related to bone, either in systemic illness or in traumatic injuries. Part of these treatments evolved from clinical to the basic science of understanding the underlying cellular mechanisms, and the recent is evolving by the "natural" approach from understanding the biology of osteoblast toward the clinical application, e.g., in vitro generation of viable bone for the clinical use as an autologous bone graft (Table 10.1).

Still, there is a significant gap of knowledge for understanding the enigma of why different modalities and parameters of biophysical stimulation of osteoblast cause similar results of induction of bone regenerations. There is a crucial question if these phenomena are induced by an identical chain of interrelated cellular pathways or by unrelated mechanisms for different types of externally applied energy. Is there is a possibility of a "Uniform theory" of biophysical stimulation of osteoblasts? This question is crucial for optimizing the clinical use of the biophysical stimulation of bone cells for bone regeneration. The resolution of this enigma will have a significant impact on the future of the treatment of disabilities related to bone.

N. Rosenberg, *Biophysical Osteoblast Stimulation for Bone Grafting and Regeneration*, https://doi.org/10.1007/978-3-031-06920-8_10

Table 10.1 Summary of biophysical stimulation modalities that induce bone cells for matrix elaboration

Method	Energy type	Frequency (Hz)	Current clinical use	Foreseen clinical use	Medical condition treatment
Distraction osteogenesis	Mechanical	0	+		Bone deformities
High-frequency mechanical vibration	Mechanical	10–70	+(1)	+(2,3)	1. Osteoporosis 2. Bone fractures 3. In vitro generated autologous bone graft
Ultrasound	Mechanical	>20,000	+		Bone fractures
Pulsed electromagnetic fields	Electromagnetic	10–20	+		Bone fractures
Radiofrequency pulsed electromagnetic fields	Electromagnetic	2100×10^3	–	+	Bone fractures
Pulsed visual light	Electromagnetic	40	–	+	Bone fractures

The manufacturer's authorised representative in the EU is Springer Nature Customer Service Centre GmbH, Europaplatz 3, 69115 Heidelberg, Germany. If you have any concerns regarding our products, please contact ProductSafety@springernature.com

Printed and bound by CPI Group (UK) Ltd, Croydon, CR0 4YY
15/07/2026
02167643-0001